P9-CAL-456

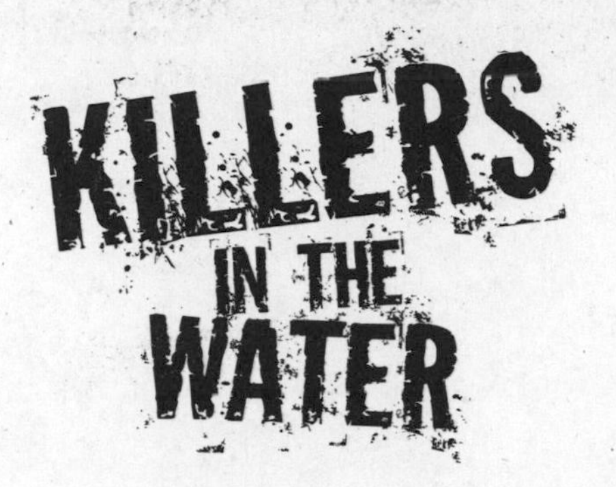
KILLERS
IN THE
WATER

SUE BLACKHALL

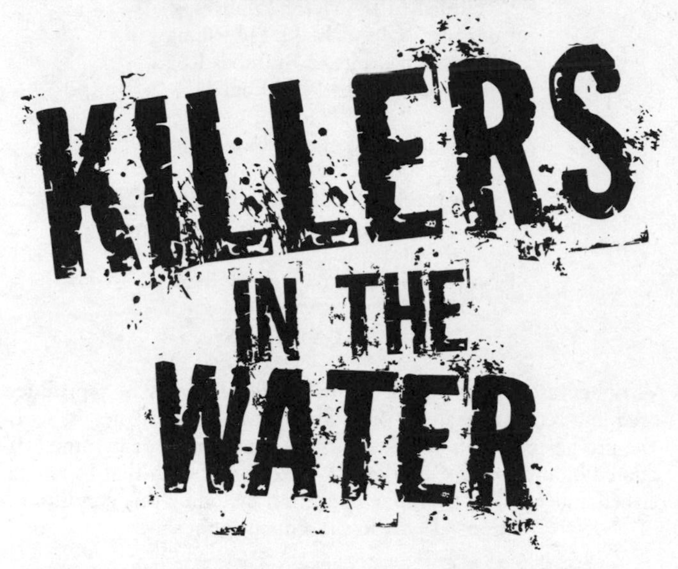

THE NEW SUPERSHARKS TERRORISING THE WORLD'S OCEANS

Published by Metro Publishing
an imprint of John Blake Publishing Ltd
3 Bramber Court, 2 Bramber Road,
London W14 9PB, England

www.johnblakepublishing.co.uk

www.facebook.com/Johnblakepub facebook
twitter.com/johnblakepub twitter

First published in paperback in 2012

ISBN: 978-1-85782-669-2

British Library Cataloguing-in-Publication Data:

A catalogue record for this book is available from the British Library.

Design by www.envydesign.co.uk

Printed in Great Britain by CPI Group (UK) Ltd

1 3 5 7 9 10 8 6 4 2

Papers used by John Blake Publishing are natural, recyclable products made from wood grown in sustainable forests. The manufacturing processes conform to the environmental regulations of the country of origin.

CONTENTS

INTRODUCTION

> 'Shark attack is probably the most-feared natural danger to man, surpassing even hurricanes, tornadoes and earthquakes in the minds of most beach users and sailors.'
>
> Dr. George Burgess, director of the Florida Program for Shark Research and curator of the International Shark Attack File at the Florida Museum of Natural History

It is impossible to hear the word 'shark' without an accompanying feeling of terror: its jaws are inescapable, its ability for savagery the stuff of which nightmares are made. The world was alerted to the mayhem caused by sharks in the 1975 film classic *Jaws*. That was fiction of course. But the nightmare became fact in 2010 when a 'serial-attacker' shark caused death and terror at the holiday resort of Sharm El Sheikh, Egypt. Just six months later, in August 2011, the whole horror was re-ignited with the savage death of British man Ian Redmond while on honeymoon in the Seychelles. But attacks by some of the world's biggest beasts are worldwide –

and have been occurring for hundreds of years. Only over the last ten years has there been a huge increase in shark encounters. With more people wanting to get up close and personal with sharks and with dive companies more than willing to fulfil this desire, an even bigger increase is expected over the next decade. In 2011 there were 120 reported attacks, which included 20 deaths.

Many theories have been put forward as to why sharks attack including changes in sea conditions, over-fishing and familiarity with humans as more and more of us take to the waters. But there are some pretty startling revelations, too – some quite spooky – as to why the attacks take place. With a mystique that equals any science-fiction plot, the greatest number of attacks occurs during new moons. Experts believe this is because the new moon influences the movements and reproductive patterns of fish and sharks have developed the sense of knowing just what these changes are – and at exactly what time shoals are available for hoovering up. If you are in the water at the time, a shark will not differentiate between you and its fishy prey. It is the shark's massive bulk and ferocious teeth that does the damage, of course. Horrendous injuries are caused simply because a human is tiny prey compared to thc predator's size.

You are also more likely to be attacked if wearing a black-and-white swimsuit, simply because the combination gives a shark a good colour contrast on which to feast its eyes. Oh, and don't go into shallow water on a Sunday. In short, research by shark expert Dr. George Burgess shows that attacks are most likely to occur on a Sunday, in less than 2m (6.5ft) of water and during a new moon. More men than women are attacked, quite simply because more men take to the water. In most cases the shark will bite only once – either to see it if it likes the taste, to defend itself or simply in an act of aggression.

Sharks learn pretty quickly when it comes to finding food. In January 2012, a Tiger shark caught in the Gulf of Mexico coughed up feathers. It was just one that had caught onto the fact that birds disorientated by lights from oil rigs are easy pickings but then, having been around for over three million years, sharks have become pretty smart.

This book is a round-up of the major shark attacks throughout the world over the last ten years or so. It is impossible to mention them all as the number amounts to literally hundreds. I have tried to include as much detail as possible for each attack, but sometimes recorded information is a little vague, with some victims not identified or dates varying. This is understandable with so many attacks occurring at so many different locations.

I have also included shark facts and the odd extraordinary shark tale – sometimes these monstrous beasts create their own monstrous stories.

The experts tell us we have more chance of being knocked over by a car than being attacked by a shark, but read some of these shocking shark stories first and make up your own mind…

CHAPTER 1

THE RED SEA SERIAL KILLER

'I BEGAN TO REALISE I WAS BEING EATEN ALIVE...'

Egypt's tourist hot spot the Red Sea rightly deserved its name at the end of 2010 for the crystal-clear waters, a haven for divers because of the prolific sea life, were indeed turning red – with the blood of those maimed and killed by a shark.

But this was no one-off attack. It was if the beast was lying in wait for its prey; its taste for blood sharpened, its natural instincts enflamed to such a degree that human flesh and bone were literally easy meat.

Death and devastation were the only words to describe the events that occurred at the holiday resort of Sharm El Sheikh during the serial-killer shark's reign. The attacks left one woman dead and four other people horrifically injured – and memories equally scarring. So desperate were the authorities to bring an end to the nightmare that chaos and confusion abounded – with the predator proving as slippery as its stealthy movements below the water. Its deadly attacks were killing off the beach tourism that contributes a massive percentage towards the £7.8bn ($12bn) or so reaped by Egypt's travel industry every year.

The shark seized its first victim around 2.40pm on Tuesday, 30 November 2010 at Tiran Beach. Russian woman Olga Martsinko, 48, lost an arm and was horrifically mauled as she snorkelled with her daughter (Elena, 21) in shallow waters close to the beach with the darkly ironical name of Shark's Bay. The first time Olga realised she was not alone in the water was when her outstretched hand touched a solid bulk. Warnings of 'Shark! Shark!' were screamed in Russian by fellow holidaymakers. An Oceanic Whitetip – the species described by famous underwater explorer Jacques Cousteau as 'the most dangerous of all sharks' – was on the attack. Olga later gave a graphic account of her horrific ordeal.

'At first I thought it was a dolphin, and then this black fin; I saw it right in front of me, this black fin. When it caught me by the arm I felt three rows of teeth. I felt a sharp pain as it came up, sank its teeth into my arm and began to wag me around. I knew there were predators in the sea but I never expected to meet a monster face to face. Like a spark, a clear thought flashed through my mind that it was a shark. It went under me and bit my buttock and tore it off; tore one of my buttocks right off. It tried to pull me down with it into the sea and I saw the huge jaws and sharp fin beside me. The shark let me go for a second and I swam away. But it came back for more, biting me again and again from behind. If I had not had my flippers on, it would have taken off both my legs. I began to realise that I was being eaten alive; that I might not be able to reach the pontoon. I immediately realised I could die, right then.'

By the time Olga made it to the pontoon to be dragged to

safety by shocked tourists, she had lost a chunk of her right thigh and buttock; her hand and arm were also missing. The waters were red with her blood and still more poured from her appalling injuries.

Olga's daughter gave her own emotional account: 'We were only 30 feet (9m) away from the jetty and could easily see the other people on the beach. My mother suddenly disappeared under the water and I saw the shark swirling her down. When she came up again, the shark let go and she shouted "*Spasite, akula!*" ("Help me, shark!") If she had not swum fast with her flippers, she would have lost her legs as it chased her to the jetty.'

Olga was flown to Cairo's Nasser Institute Hospital, where the surgeon on duty, Dr. Mohamed Dahi, was shocked at the severity of her injuries – 'I walked into the emergency room and when I saw the victim, I found her arm was amputated up to the elbow. I saw the wound at the back of her. It was about 40cm (16in) by 50cm (20in).'

Just minutes after the attack on Olga there was a second one. At 2.55pm, another Russian woman, Lyudmila Stolyarova, was swimming close to the shore and the pontoon, and well within the buoy area. She felt something brush against her and assumed it was a diver coming to the surface of the sea. But it was a shark: 'It swam between me and the sea shore, so you see I had nowhere to go,' said Lyudmila. 'It started to swim in circles around me, in circles, and right then it started – started to attack me. It bit my arm off. It bit it off! I lifted my arm up in the air and started screaming "Help, shark!" Only when they started hauling me onto a boat did I realise my leg was bitten off.'

Meanwhile, her husband Vladimir, 72, had been forced to

look on helplessly. 'I looked up and there – oh no! People were carrying my wife on a sun bed and the ambulance was waiting. It was such a horror. What I saw there when she lifted her arm, I won't describe all the horrors, the torn wounds.' The special holiday break the couple had been enjoying to celebrate Lyudmila's 70th birthday ended in a nightmare, with Vladimir's pleas to his wife not to go into the water because he felt uneasy proving justified. Diving instructor Hassan Salem, who had been on the dive, told how he was circled by the shark before it attacked the victims: 'I was able to scare the shark away by blowing bubbles in its face, but then saw it swim to a woman and bite her legs,' he said.

Lyudmila, too, was a victim of an Oceanic Whitetip and was also rushed to the Cairo hospital. Again, Dr. Dahi was confronted with the devastation caused by a shark on the human body: 'She had an amputated arm and an amputated leg. At our hospital we have experienced injuries caused by sharks. They were mainly caused by small sharks – not like this. This time it looks to be a large shark, I feel there is a problem in the sea.'

The next day, Wednesday, 1 December at 10.55am, Ukrainian snorkeller Viktor Koliy, 46, was attacked within minutes of jumping from the pontoon in Shark's Bay. But it was this attack that made the authorities realise that not one, but two species of shark were preying on humans. This time the attacker was a Mako shark. Yet, like the Oceanic Whitetip, it was far from its deep-sea home and had strayed into shallow waters. Koliy's arm was torn and he was pulled from the water, the haemorrhaging from his injuries turning the sea red again.

Just five minutes later, 54-year-old Russian marine captain

Yevgeniy Trishkin became the next victim to be savaged in a shark's jaws; not those of the Mako, but an Oceanic Whitetip. His left arm was torn off from below the elbow and his other hand mauled as he swam near Naama Bay on the third day of his holiday. Like the other victims, Trishkin had been swimming in the 'safe' area marked for swimming. Children had been playing in the water and there were no warning signs about the threat of sharks.

Afterwards Trishkin recalled: 'At the end of the pontoon there was a sharp depth increase and the colour of the sea gets a much darker blue, almost black. I swam into this and by the time I saw the shark, it was too late. It came from the deep and was the same black colour as the sea; it bit and started chewing my left arm. I hit it with my right arm while it continued to eat the left away. I hit its nose several times and for a second it opened up its jaws, but only to bite my other hand. It was a huge creature. The people on the jetty heard my cries and came to rescue me just as I was losing consciousness. They dragged me out of the water.'

His ordeal was witnessed first hand by British holidaymakers Jim and Joanna Farr, who had been snorkelling over the coral reef. Before swimming away in terror, they quickly took photographs of the attack, which identified the Oceanic Whitetip shark as the one responsible for previous attacks because it had the same chunk missing from its dorsal fin. Jim, 58, had actually been bumped by the shark before it made its way to its victim. He remembered: 'We were in the water, just swimming along quite casually when we came across an area which was a diving pontoon. Below us was the coral reef and three divers or four divers down there; you could see all the bubbles, but it was as clear as anything. It was wonderful, fish galore. Two guys jumped off the pontoon just in front of us and they swam past me. As they swam past, I felt this bang on my back – I thought they had jumped off the pontoon on top of me. But I looked up and could see the pontoon was about ten feet away from me.'

Jim and his wife Joanna, 42, were dragged to safety in a boat. Joanna takes up the story: 'The snorkel guide took his snorkel out and shouted "Shark, shark! Get out the water now! We were about twenty or thirty metres from the boat and we literally just had to swim for it, knowing there was a shark in the water. There was screaming and shouting; people just went wild. Everyone was in shock. You could see blood pouring through the pontoon, staining the water red. The guy next to him had his head in his hands because he had just witnessed it all. People were screaming and they were still pulling people out of the water. Some were running along the platform to the safety of the shore and people on the beach had run back in terror, standing 10m (33ft) from the shoreline to get as far from the water as they could. There was total panic. Our guide told us he had seen a 3.5m- (11ft) long shark in the water and pointed to the Oceanic Whitetip in my fish guide. They took us away from the scene as quickly as possible and tried to convince us it was probably someone who had cut themselves on the coral but we knew it was a shark and nobody wanted to get back in the water.'

Jim adds: 'I could see the guy being pulled up onto the pontoon. There was blood everywhere, blood squirting everywhere – it was like a war scene.' The couple literally had to swim for their lives. 'It was only when we got back [that] we heard the shark had attacked several swimmers and realised what a close shave we had had,' said Joanna. Among the photographs they took were some that showed blood pouring off the pontoon as Trishkin was pulled to safety.

Local diving instructor Marcus Maurer helped with the rescue. 'There was a lot of blood. I saw the shark stay there for a while and we started the rescue situation. A lot of people had been carrying the man up, so we gave him oxygen support and we brought him directly to hospital. Usually we do not see sharks. We have lots of people during the year snorkelling, swimming and diving at this place. You do not see sharks because they are shy animals.'

Mohamed Rashad, a barman at the nearby al-Bahr beach restaurant was one of those who could only look on in horror as those in the water attempted to flee to safety. 'The sea went red!' he said. British holidaymaker Nina Dydzinski, 46, from Wigan, Lancashire was relaxing on the beach when she heard the shouts – 'I had just come in from the water, where I'd been snorkelling close to the beach. I heard a man shouting. Everyone panicked.' Her husband, Jarostaw, 49, was lucky to escape. He was in the water when the hysterical cries rang through the air, but managed to scramble to safety. 'Everyone was just trying to climb onto the jetty,' he said.

No wonder tour operators reported a significant drop in the number of Russian tourists making their way to Egypt. 'There is no doubt that the news coverage of the shark attacks in Sharm El Sheikh has scared tourists away,' admitted Irina Sivenko of Moscow's TanTour tour agency, before adding that while most tourists were refusing to go to Sharm El Sheikh, others had actually contacted the company to see if package holidays there were now cheaper.

Despite determined efforts to stress the rarity of shark attacks in the Egyptian waters, this was no consolation for holidaymakers or those endeavouring to bring an end to it all. 'We are monitoring the situation very closely and working together with all authorities to ensure the safety of all members and visitors in the Red Sea,' said Hesham Gabr, chairman of Egypt's Chamber of Diving and Watersports (CDWS), 'Our thoughts are with the victims and their families.'

No one could have imagined that worse was yet to come.

Desperate not to lose vital tourist trade – up to five million holidaymakers every year – the authorities knew they must hunt down the killer. But was it one main predator or more? Some had witnessed the Oceanic Whitetip in action; others reported seeing the Mako shark. The CDWS stated that it was 'working continuously with all the relevant authorities and shark experts to try to resolve this situation in the most appropriate and safe way for all concerned.' Furthermore, it was calling on the help of experts to 'form an advisory team on the best course of action' following the Naama Bay incident. In other words, it was time to bring in the heavy mob. Three shark experts from America were flown in to Sharm El Sheikh – Dr. George Burgess, director of the Florida Program for Shark Research and curator of the International Shark Attack File at the Florida Museum of Natural History; Marie Levine, head of the Shark Research Institute at Princeton, and Ralph Collier of the Shark Research Committee.

'When you have these many shark attacks in such a short period of time, there must have been something to incite it,' pronounced Dr. Burgess. 'It does not conform to normal shark behaviour in the least bit.' Collier said: 'I have been working in the field of shark interaction since 1963 and this is the first time I have ever seen injuries this severe and this localised as far as the area of the body that was bitten is concerned. We always hear that sharks like human flesh, but of course that is not true. Sharks don't like humans – it's as simple as that.' Back in the US, a fourth expert – shark behaviourist Erich Ritter – was on hand to advise from his research centre.

'My job was to figure out why,' he said. 'Every wound tells a story. Sharks' teeth are like fingerprints. They are identifiable to a specific species based on the shape and the function of the tooth. Mako sharks have sharp, pointed teeth which slash through their prey, whereas Oceanic Whitetips have serrated teeth which leave a straight cut in their victims. What this told me was that most wounds were inflicted by an Oceanic Whitetip, but one person was actually bitten by a Mako shark.' All of which surprised Ritter. 'Humans are not the normal prey of sharks – we are terrestrial animals, we don't live in the ocean. Sharks feed on things they see and live with every day in the marine environment. We are not on the menu. If they are showing up in shallow waters, then there must be a very powerful trigger.'

It was a view shared by diving instructor Marcus Maurer, who said that although the powerful Oceanic Whitetip was indigenous to the Red Sea, thousands of divers had encountered them without any problems: 'I have now had more than 3,000 dives and I have never had a problem with them. This is a really unusual event. I think the chance of dying by an aeroplane accident is much bigger than getting involved in a shark attack. The instructors and divers who come here are actually looking for sharks because we love them. And that is really, really, an excellent experience for everybody.'

Yet another shark expert – Samuel Gruber, head of Miami's world-famous Bimini Biological Field Station – described the attacks as 'unprecedented'. He said: 'A shark in one day bit more than one person. In all my years reading about sharks and writing about them, you never hear about sharks biting more than one person, then for it to happen the next day is almost like a *Jaws*' scenario. Finding the shark is pretty much a crapshoot – it's like trying to find a needle in a haystack.'

Gruber said such frenzied feeding is normally reserved for shipwreck survivors. The most infamous took place during World War II when the *Nova Scotia* – a steamship carrying around 1,000 people – was sunk near South Africa by a German submarine. With only 192 survivors, many deaths were attributed to the Whitetip. Packs of sharks moved in for the kill and those on rafts could only look on in horror as their fellow passengers, desperately thrashing about in the water, were eaten alive. Sergeant Lorenzo Bucci recalled: 'A lone swimmer would appear, then suddenly throw his arms in the air, scream and disappear. Soon after, a reddish blob would colour the water.' Later, around 120 corpses washed up on Durban's beaches. And on 30 July 1945, the *USS Indianapolis* was torpedoed, with many of the 800 sailors on board succumbing to shark attacks, as well as exposure. In fact, the Oceanic Whitetip is responsible for more fatal attacks on humans than any other species combined. There were five such recorded attacks in 2009.

Meanwhile, back at the Rea Sea and events of 2011, some expressed cynicism about the intervention of high-profile experts. One dive-centre owner observed: 'Why did they need to import all these specialists merely to come up with the same explanations that we all had from day one?'

More help came from a Swedish vessel surveying the waters around the resort to try and track any shark movements. The authorities closed all Sharm El Sheikh's beaches and all diving and watersports activities were suspended. Now the once tourist-thronged beaches were virtually deserted, although perhaps surprisingly consent was still given for experienced divers to enter the Red Sea. But even venturing to 'safe' locations ended in fear for some intent on hitting the waters. London financier Nick Treadwell visited the national marine park of Ram Mohammed, where he set off on a boat trip in a party of ten snorkellers. Treadwell was one of two who decided to go scuba diving. The morning's dive was idyllic, but the second one was certainly not. Treadwell was down about 14m (46ft) when his instructor noticed a stray piece of equipment below them and dived for it, before flinching in shock on his way back up. Above the scuba divers and below the snorkellers a large shark, more than 2m (6.5ft) long, slowly encircled the group. Treadwell later told the *Guardian* newspaper that after it was brought to his attention, he went 'calm – very calm' as he watched the instructor freedive down a few metres and then begin corkscrewing to the surface, blowing bubbles, in an attempt to scare the shark away: 'He went to the top and shouted, "Shark, shark, shark – everyone get to the reef!" Everyone started swimming as fast as they could because the reef was too shallow for the shark, so it would be a safe place. But there was an older lady, probably in her late 60s, who was slightly hard of hearing and she was delayed.

The shark started coming towards her, and she ended up kicking it in the face a couple of times and using her underwater camera to whack it over the head. She got away, but she had cuts all over her legs. I don't know whether the shark had actually bitten her, but they looked like lacerations – almost like injuries from where she'd kicked the shark in the teeth.'

Dr. Mohammed Salem, marine biologist and director of South Sinai National Park, stressed that it was wrong for people to assume sites like Ras Mohammed were safe: 'It is a misunderstanding when people think that the words "National Park" mean that it is a place that is absolutely protected from human activities. I tell them this is wrong.' In fact Salem was one of the few to be open about the reality of shark attacks in the Red Sea, admitting that between 1996 and 2009 there were 12 attacks recorded, the latest bringing the number to 17. However, he added, 'This is a very small number compared to other attacks in other countries like the USA and Australia. Also, regarding the very high number of snorkellers and swimmers coming every year to the area, we find this number is slight.' He added that the recent attacks had occurred from north Naama Bay to Ras Nasrani, just by Shark's Bay, and 'had our attention because naturally, sharks fear human contact because they don't know what they are so [they] are careful when they come close to them. The other observation was realising that the attacks were in the day while naturally, sharks hunt only during the period between sunset and dawn. The question is what made these sharks in this specific area change their behaviour and attack.'

Meanwhile, the hunt for the killer – or *killers* – went on.

Beliefs about the shark attacks still differed, though, with one senior government official stating: 'It is clear now that we're dealing with multiple sharks and undoubtedly at least one of them is still out there in the water.' Boats and divers were dispatched to track it – or *them* – down. Desperate to allay fears, make their waters safe and preserve a highly lucrative tourist industry, the authorities knew they must act fast. But it was all about to go horribly wrong.

In a statement, the Egyptian environment minister announced that the shark had been captured and was being held at Ras Mohammed. Two other sharks – a Mako (the smaller, shyer species rarely seen in the Red Sea) and an Oceanic Whitetip – were slaughtered and dissected for examination of their stomachs. The examination of the Mako confirmed it was indeed responsible for the attack on Viktor Koliy: an anomaly in its teeth, probably after an injury from a hook, matched their wounds. Film of the catch showed the Oceanic Whitetip being hauled by its eye sockets onto a small fishing boat, its dead body cut and bloodied, belly slashed. Both creatures had been caught after reaching for bait thrown into the sea by their hunters. A government statement issued in a bid to bring about calm said the Oceanic Whitetip shark was similar to the one photographed by a diver minutes before the first attack. Others insisted it was not the shark and they would be proved right because the Oceanic Whitetip that had been captured and killed did not have the damage to its fin as described by witnesses to the attacks. Also, the wounds inflicted on victims did not match its teeth. What everyone did agree on was that such species of shark normally stay in deep waters, a habit now broken with devastating results. Biologist Dr. Elke Bojanowski said: 'I was very surprised that the first shark they caught was a Mako shark. I have been working in the Red Sea for seven years and I have never seen a Mako shark underwater in an area like that, close to the reef. Usually they are open-water sharks and do not come so close to the reef.' In all, some 40 diving instructors volunteered to trawl the waters, but no further sharks were found nearby.

Meanwhile, there was growing criticism of the way the authorities were handling the terror that stalked the waters: warning signs should have been erected after the first attacks, people ought to have been alerted to the danger lurking in the waters and more should have been done to publicise such a monumental occurrence. Furthermore, simply tracking and killing any shark to be found was not enough: it was a random approach and no one was convinced the real killer had been stopped. 'I have always said that there was no way this could be the work of a single animal,' said Amr Aboulfatah, former chairman of the South Sinai Association for Diving and Marine Activities, and the owner of a large local dive centre. 'You've got more chance of winning the jackpot in Las Vegas than you do of identifying and then capturing a single shark and thus solving the problem.' The CDWS agreed with him, saying that it did not 'in any way condone the random killing of sharks.' Some conservation groups said that the authorities had originally promised to relocate the rogue shark – or sharks – to the Gulf of Suez, but had instead sent a 12-man team to capture and kill two which are listed as vulnerable on the International Union of the Conservation of Nature's (IUCN) Red List of Threatened Species.

Even more controversially, Amr Ali, director of the

Hurghada Environmental Protection and Conservation Association (HEPCA), said he had received photographs of ten sharks killed by the authorities. The group accused Egyptian officials of attempting to 'wipe out local shark populations'. Dr. George Burgess declared a shark hunt to be a waste of time as his team had ruled out the existence of a so-called rogue shark that was acting like a 'deranged human being taking lives.' Meanwhile, Dr. Bojanowski said that the random killing of sharks did not help anyone: 'You put out baited hooks and then a random shark that might have just entered the area and maybe was not even there the day before is just grabbing that baited hook. You could also be attracting sharks to the area. It is not helping the problem. They are very, very self-confident; very, very curious. A shark will actually approach you and check you out – you could probably best describe them as bold.'

But as far as the authorities were concerned the attacking shark had been caught. The waters were therefore re-opened and declared safe but just in case, they drew up instructions for local hotels and diving centres. 'In line with these instructions, hotels and diving centres will have to appoint special staff, who will permanently supervise swimming areas and, if needed, report sharks approaching beaches,' announced the Ecology Ministry. The instructions also meant people had to be evacuated from insecure areas if a shark sighting was reported and a shark hunt subsequently launched.

Yet it was no wonder that no one felt reassured – the serial killer of the deep was causing terror equal to that of any disaster movie. And so many knew the killer was still out there. Then came a horrific climax to the attacks.

On 5 December a 70-year-old German woman, Renate

Sieffert from Markdof, died in the shark's jaws. She was already dead when pulled from the water. Her partner, Rudi, could only look on from the beach, his screams and tears mixed with stunned horror. Renate's death came just 24 hours after the Red Sea was pronounced free of any threat of sharks. Dr. Bojanowski commented: 'We opened the beaches and everything was normal – there was no shark activity at all that day.' Renate Sieffert had been a regular guest at the resort's luxury Hyatt Regency hotel. Diving instructor Ehat Abd Elrahman was with divers completing their open-water qualification when he spotted an Oceanic Whitetip close to the shore and raised the alarm: 'It was very big – I had never seen one so big. The shark did not pay any attention to us. He did not have any problem with divers but for sure there was something going on, on the surface. I raised my hand to signal that there were sharks in the water, but by then things were already happening.'

Tourists described in graphic detail how the waters went red as the victim snorkelled just 20m (66ft) from the shore. Sieffert's arm was severed and she died within minutes. Ellen Barnes, 31, a British tourist who was in the water at the time, recalled the scene vividly: 'I looked behind me and there was this woman thrashing about and screaming for help. The water was full of blood – it was horrific. The shark kept coming up and taking bites out of her and then coming back for more; for another bite. The water was churning like I was in a washing machine; I was being thrown around in the blood. The shark was thrashing and tearing at this poor woman and I could barely keep my head above the water.'

Ms Barnes, from Horsham, Sussex, described the lifeguards

at the beach as 'useless and petrified'. Her partner, Gary Light, 32, shouted to them to get the swimmers out of the water but noted they just stood there: 'I could see the shark taking bites, and going back and attacking this woman. I was trying to get Ellen and everyone out of the sea. It was ghastly, horrible, like something out of a horror film.'

Ms Barnes went on to criticise the authorities for assuring tourists that the sea was safe: 'We were kept very much in the dark about it and we were all promised that the sea was safe. We were told that the sharks would not come over the coral as it would scratch their bellies, so that's why I was out there snorkelling. It is a big shock. The worst thing is that the lifeguards promised there were no more sharks in the sea. I think it is such a shame that the Egyptian Government felt confident enough to send out faxes to all the hotels to say it was safe when obviously it was not.'

Jochen Van Lysebettens, manager of the Red Sea Diving College at the resort, told Sky News: 'The woman was just swimming to stay in shape. Suddenly there was a scream of "Help!" and a lot of violence in the water. The lifeguard got her on the reef and he noticed she was severely wounded.'

Another swimmer, Inna Koval, described the attack: 'I was snorkelling a couple of minutes away before the incident. It happened just 100m [328ft] away from me. It came so close. I heard a man who started to scream very loudly. All I could understand was that something was wrong and the word "shark". I saw the tussle in the water and a tail of a shark for a second above the water. Many people were still swimming. It happened so suddenly that they struggled to get away very quickly. It was so close to the beach where tourists were allowed to swim. No one expected this because the waters are not deep enough and there is coral everywhere. A speed boat which was cruising around came by immediately and circled the victim and the shark, but it couldn't do anything else. The attack went on for about seven seconds but it was long enough to get its victim.'

Yet another casualty and this time a fatal one for Dr. Dahi. 'She was already dead when she came to the hospital,' he recalled. 'She had an amputated right arm and highly amputated right leg. There was a large wound to the back also – I have never seen this injury before.'

It was enough for the Chamber of Diving and Water Sports (CDWS) to send an urgent message to its members in Sharm El Sheikh ordering them to clear the water. It read: 'Following reports of another incident in Middle Garden local reef, CDWS is calling for all its members in Sharm el-Sheikh to stop any snorkelling activities happening from any boats or shore. Please tell all your boats to immediately recall any snorkellers who may be in the water.' Group chairman Hesham Gabr added: 'We are busy dealing with the crisis. I can confirm that a German woman was injured and she passed away.'

A fuller statement followed: 'CDWS announced this evening that all diving and watersports have been suspended along the Sharm el-Sheikh coastline tomorrow (Monday 6 December 2010). The suspension comes following a 4th incident in less than one week involving a shark attack on a tourist. Today's event took place off the beach in front of the Hyatt Hotel, Naama Bay. Unfortunately, the 70-year-old German woman did not survive. Last week, similar attacks took place involving Ukrainian and Russian snorkellers. These incidents led to severe injuries, but no loss of life.

CDWS is the regulatory body for diving and water sports in Egypt and would like to emphasise that such attacks are extremely rare and this kind of shark behaviour is causing disbelief amongst the Red Sea diving community.'

The shark attacks in the Red Sea also prompted an official safety warning from the British Foreign & Commonwealth Office (FCO). It is a warning which still remains on its information website:

> Safety and Security – Adventure Travel
> Before undertaking any adventure activity ensure that your travel insurance covers you for the activity.
>
> If you are considering diving or snorkelling in any of the Red Sea resorts be aware that safety standards of diving operators can vary considerably. A basic rule is never to dive or snorkel unaccompanied. Where possible make any bookings through your tour representative. Unusually cheap operators may not provide adequate safety and insurance standards. Ensure that your travel insurance covers you fully before you dive. Diving beyond the depth limit of your insurance policy will invalidate your cover.
>
> The Egyptian Chamber of Diving and Water Sports (CDWS) website provides further details and regular updates on diving conditions in Sharm el-Sheikh, including advice following a number of shark attacks on 30 November, 1 December and 5 December in which a tourist died and three others were injured. Shark attacks of any kind are very unusual in the Red Sea but we

advise that you monitor updates issued by the local authorities and your tour operator.

Ensure that your travel insurance covers you fully before you dive. Diving beyond the depth limit of your insurance policy will invalidate your cover. You should also ensure that your travel insurance, or that of the tour or dive company, provides adequate cover for the costs involved in any air/sea rescue if you are lost at sea. The current fee can exceed US$4000 per hour. The Egyptian authorities will only undertake air/sea rescue operations on receipt of a guarantee of payment. The British Embassy is unable to provide this initial guarantee, but does facilitate communication between insurance companies and the Egyptian authorities.

The German Embassy in Cairo, which was involved in returning Renate Sieffert's body home, said it would not issue a warning to German tourists about the threat of sharks, stating: 'We look at this as an unfortunate, sad, but classic accident. It is obvious that those who go swimming in the Red Sea should be careful – especially after the incidents of the past few days.' Salem Saleh, director of Sharm El Sheikh's tourist authority, had to admit: 'We did make some efforts last week, but I think we failed.' Gen el-Edkawy of the South Sinai Government reported no tourists had cancelled trips to the resort following the death. Bent on proving all was well, he donned a wetsuit and jumped into the water just yards away from where Renate Sieffert was attacked. After 20 minutes he emerged, pulled his snorkel mask to one side and announced: 'I saw a lot of beautiful marine life – it was wonderful, everything is wonderful! This city is a gift from God and I'm sure everything is safe.'

Egypt's tourism minister Zuhair Garana now declared the Red Sea safe: 'I cannot say that deep waters are completely secure but shallow waters are 100 per cent secure. Diving is being allowed. We are advised that sharks will not attack divers.' South Sinai governor Mohammed Shosha lamely proffered: 'We did catch the sharks – there is another shark.'

That other shark was a female Oceanic Whitetip – and the very one witnessed by those in the shallow waters moments before Renate Sieffert was fatally attacked and before the first victim, Olga Martsinko was savaged. While an 'innocent' member of the species had been slaughtered, the real killer – complete with the distinctive fin damage identified by so many and which should have led to a much earlier ceasing of its reign – had been left at large. 'There were underwater pictures and videos of the shark. She had very distinct markings in her tail mainly. There was a clear indentation at the upper edge of the tail. That was a very rare notch for a shark to have, so she was easy to identify,' said Dr. Bojanowski.

It is perhaps understandable that no one knew exactly what to do about the predator. The last shark fatality in Egypt – and the first for five years – had been in January of that year but it occurred just south of Marsa Alam, a remote diving area a long way from Sharm El Sheikh, where such attacks are almost unheard of. Indeed, worldwide, statistics proved just how extraordinary the Red Sea Attacks were. According to the International Shark File, which has a global shark database, there had only been nine attacks on humans by Oceanic Whitetips since records began in 1580 and only one of those had been fatal. One diving centre owner claimed the local shark attacks were the first around Sharm El Sheikh for 15 years, although other locals said there had been an attack two years ago and divers claimed they had alerted the authorities to the presence of sharks in the last few months. Ezat Ezat of the Wave Dive Centre said that people want to dive among the sharks: 'They get angry if there are none around. It's good for us to go and see the sharks, but not so good when the sharks come and see us.'

Just a few weeks before the shark attacks in Egypt, Dr. Avi Baranes, a scientist at the Interuniversity Institute for Marine Sciences, presented a report to the Israel Academy of Sciences and Humanities summing up 30 years of research on Red Sea sharks. In an interview with Haaretz.com, Baranes said: 'Sharks are actually quite nice. It's true that a violent confrontation between swimmers and sharks is an exceptional event. I think there have only been two cases in all our history: with a British English soldier in 1946 and an attack in Eilat in 1974. Unfortunately, the film *Jaws* gave sharks a bad name, and unjustifiably so. Many more people are eaten by dogs than by sharks, and this is because sharks are uniquely attached to their own particular food. So, when there are incidents like this, we have to look for a cause. Sharks attack people when they invade their territory, and then their reaction is aggressive, as it is when they are unable to obtain their natural food.'

The marine expert added that a Mako shark attacked a German tourist in 1974 at the popular diving resort of Eilat (the shark was caught the next day and it was found to have a spinal problem and could not swim fast). Said Baranes: 'This was a shark that usually fed on tuna and it simply could not obtain its regular food. In Sinai I know of cases where sharks simply bit the legs of Bedouin fishermen who were standing on reefs. This is the response to an invasion by humans into the shark's living space, and this can also happen when divers enter their territory. I suggest we not be afraid of sharks. We must respect them, and we can look at them in the water. I also suggest that we refrain from trying to attack them because they will respond and they have the strength and the means to defend themselves. Just enjoy the view. There is nothing more beautiful than a shark swimming in the sea.'

It was now obvious the authorities had to do more. Again the beaches were closed and signs erected with the warning: 'As per the latest instructions by the South Sinai Government, please avoid swimming in deep water as there are threats related to sharks.' Sales of snorkel masks and flippers from the beachside shops dropped. The once-humorous T-shirts with a picture of a shark and the words 'How 'Bout Lunch?' were no longer funny.

For some, such actions were not enough after the fatality and other attacks – the threats should have been heeded earlier. Accompanied by his wife, British man Terry Collins was in a party of snorkellers the day before the German tourist was killed and the group had been menaced by an aggressive Oceanic Whitetip shark. 'It was about 3m [10ft] long. I was about 10m [33ft] behind everyone else. I saw it come out of the depths and it went towards our leader. It circled him and began circling the group. It was deep grey and was that close I could see electric blue fish swimming in front of it. It was circling lazily but with intent,' he said, adding that on raising his head, he saw people on the snorkel boats shouting warnings. One by one, the snorkellers had to make their way to a reef before swimming about 100m (328ft) across open water to reach the boats. 'We tried to keep the splashing down,' Collins wryly observed. His wife, Christina Stafel-Collins, recalled how the party had been forced to flee for safety after the shark circled around them: 'It was definitely an Oceanic Whitetip. We saw it so close-up. My husband is six foot and it was loads larger than him. I am so upset this woman has died – they should have shut the beaches.'

But tales of shark encounters in the Red Sea kept on coming. British grandfather Gary Young, 65, told how he was scuba diving in 10m- (33ft) deep water off an area known as Shark Reef in the Ras Mohammed National Park when he saw a 2.1m (7ft) female Oceanic Whitetip shark. It came to within 1.8m (6ft) of him. Young, of Felixstowe, Suffolk, gave his account to the *Daily Mail*: 'I was with three other divers on an hour-long dive. We were exploring the reef and looking at the fish and sea life. I looked back at my dive buddies and saw they had stopped and there was this shark coming towards us. We moved into the reef as we had been instructed to do if we saw a shark. The idea was that we would blend into the reef and any shark would be less likely to see us as a threat. She didn't seem aggressive at all. I just stayed calm and did not make any sudden movements which could have encouraged it to attack me. It occurred to me that it could have been the shark which had attacked people and it is fair to say I was a bit apprehensive. It is the closest I have got to a shark in the five years that I have been diving. I certainly would not want to get any closer. It certainly did not put me off going back into the water. You just have to treat these creatures with respect.' Remarkably unfazed, he captured the moment on his underwater camera. His pictures were sent to those trying to hunt down the predator.

Everyone, it seemed, had their own opinion on the killing waters. Richard Pierce, chairman of the UK-based Shark Trust and Shark Conservation Society, observed: 'This event is absolutely extraordinary. Since records began in the late sixteenth century there have been only nine recorded attacks on humans by an Oceanic Whitetip. It's abnormal behaviour; this shark hasn't just decided to be in the wrong place at the wrong time – there must have been a specific activity or event that brought it there. Something has brought this animal to the area and made it think dinner, and it's likely that it involves something being put in or on the water.' Like all others, Richard Pierce described the attacks as 'unprecedented,' adding: 'For either of the two species involved to make repeated attacks on humans is unheard of. They simply do not go around attacking people for fun. To see so many attacks in such a short space of time is terrifying and very difficult to understand. Behind this, there is undoubtedly some kind of human trigger.' Certainly, it was unusual to see the species as far north as Sharm El Sheikh at that time of year. The creatures may have been in the wrong place at the wrong time – but then, tragically so too were the victims.

Simon Rogerson, editor of *Dive* magazine, commented: 'If someone asked me a few weeks ago about the resort, I'd have said it was very safe. Generally, it's a really good place. It gets very crowded because it is relatively cheap for Europeans, but the coral environment is very healthy, very beautiful; the water's clear and it's sunny. Whatever is going on there is an anomaly.'

So, just what had incited the worst-ever recorded number of shark attacks that brought such devastation? Again, theories abounded. One came from Marcus Maurer, manager of the Extra Dive Centre, whose staff had been involved in rescuing the victims. He blamed the very people on whom the resort relies – tourists. 'These are open-water sharks,' he explained. 'The biggest problem is people feeding the fish. The fish are an attraction, people like to see them but if people throw food in the water, the fish come inside the reef and maybe the sharks follow the fish. They are changing the behaviour of the animals. Along the beaches, notices in several languages say "All unused food and packaging must be put in the garbage container. Food may not be eaten in the sea or within a 4ft (1.5m) perimeter around the sea". Divers know how to react in the presence of a shark. They know to stay calm, don't kick or swim fast – and don't beat thc shark.'

Maurer added that he had made more than 3,000 dives in over 13 years and had never had a problem with sharks, but snorkellers and swimmers with no experience or training simply panicked: 'Then the sharks start to hunt. People have got to learn it's not our territory, it's the territory of the animals. If we go there, we have to respect the marine life.' But it is hard to tell the thousands of children who go to Sharm El Sheikh and feed the brightly coloured fish as part of their holiday excitement that they could be attracting a killer to their feet. How are they to know that lurking so close by, sharks will pick up the pulsed vibrations sent out by shoals of fish gathering titbits and in turn attracting predators and becoming a meal themselves? And who heeds the 'Do not feed the fish' signs that are present in several different languages?

'The whole trigger is food. Nothing else makes sense,' says shark behaviourist Erich Ritter.

The unique drop in the depth of the water at this particular stretch of Red Sea beach was also given serious consideration: the reef area is very shallow around the floating pontoon area but then suddenly 'stairsteps' down, dropping off from a depth of 20m (66ft) or 30m (98ft) to more than 800m (2,624ft) – but still very close to the beach. Ralph Collier explains: 'You usually find Oceanic Whitetip sharks in waters 300m (984ft) deep or more. Because of this stairstep effect, it is not uncommon to see Oceanic Whitetips within 3m of the beach.' The investigating team also discovered the temperature of the water around Sharm El Sheikh was several degrees higher than normal for the time of year and had been over the weeks when the attacks took place. Water temperature has a direct effect on the metabolism of all species of shark. The higher the temperature, the higher the metabolic rate – meaning the shark needs more food and energy to exist. This makes them more active in hunting and increasingly aggressive in their behaviour because they want to feed. The Mako shark previously caught and dissected was found to be undernourished, implying a desperate need for food no matter where it came from. 'This animal was probably extremely hungry. It was very slim, almost emaciated. That is uncommon for Makos – their body structure is such that they are a well-built, stocky shark but this animal was not,' said Collier.

Other experts insisted that humans were again very much to blame. Divers and dive operators keen to give their clients a memorable experience were illegally feeding fish to sharks in the waters around Sharm El Sheikh also came under fire.

(In some countries you will find the illegal practices of baited dives and 'chumming' – fish blood or flesh placed in the water to attract sharks and keep divers happy.) 'This is not feeding. It is rather like one of us walking past McDonald's and sniffing the air. It attracts sharks,' said Richard Pierce. In South Africa, the Shark Concern Group has campaigned to ban shark-cage diving and chumming because it believes that it leads to sharks such as the Great Whites seeking out human company, claiming, 'risks have increased as a result of how humans are interacting with sharks.' But Pierce believes feeding sharks is a bigger risk than chumming: 'There is no scientific evidence that proves that laying chum in the water for the attraction of sharks produces conditioned behaviour. In one area of South Africa there are nine boats going out on two trips a day, chumming the hell out of anything to attract sharks for tourists. If this was producing conditioned behaviour we would expect to be seeing the same sharks there every day but we don't. The sharks have been tagged and observed, so we know. Conversely there is evidence to show that when you feed sharks, you do produce an element of conditioning. Certainly, in places where sharks are fed for tourism – in the Caribbean, for example – it has been proved that sharks have learned to associate the arrival of feed boats with being fed.'

Some disagreed, however. Shark dive operator Jim Abernethy said he believed that sharks were not inherently dangerous and went on to compare them to birds: 'Feeding the birds is an opportunity for people to get close to these animals and feed them. Birdwatchers feed birds, but every now and then a bird will bite a person by mistake.' Abernethy has his supporters. Peter Knights, executive director of

WildAid (a non-profit organisation with the aim of ending illegal wildlife trade) has dived with Abernethy several times: 'I've never been on a dive with chumming. Jim puts bait in a suspended crate, which leaves a trail of oil, not blood so it's better than chumming. The argument that sharks associate boats with food is silly because fishing boats throw bits of dead fish off the back and that industry puts more fish back that way than all dive boats combined.' Knights sees diving trips such as Abernethy's as a good method for shark tourism: 'You get people inspired and understand that sharks are not killing machines but wild animals that are sometimes unpredictable and will attack if they're confused or scared, but humans are not on their list. And if sharks get used to having people in the water, they'll realise what humans are – not food and not a threat.'

Dr. George Burgess, curator of the International Shark Attack File, disagrees with this view and does not see a difference between chumming and feeding sharks because whether it's scent or chum, sharks react primarily to olfactory signals. To prove his point, he cited the case of Markus Groh, an Austrian diver who died from a shark bite on one of Abernethy's dives in the Bahamas in 2008. Groh's 'chum bag' was said to have been grabbed by a shark: 'It doesn't matter whether it's called "chum", a "chumbag" or food, the animals go after it,' says Dr. Burgess. 'It's the equivalent of going to Africa, where a pride of lions are hanging out under a tree and dumping a bunch of T-bone steaks on the ground. That would be called dumb if you did it with lions, alligators or bears, so why do we think we can do it with the largest and most efficient predators in the ocean?' Burgess believes the dive industry tries to keep shark attacks quiet: 'Most of the cases we have on the International Shark Attack File are leaked to us. These operations are out there to make money. That's not a sin – they're trying to offer something one step above the average for people looking for thrills and the unusual. I suspect that a lot of the clients who come to these things are less naturalist-divers than people who want to be entertained.'

Simon Rogerson, editor of *Dive* magazine, agrees: 'If you were to suggest deliberately putting fishy blood and bait in the water in other countries, you'd be politely laughed off the boat.' Indeed, the practice is virtually training these creatures to expect food from outstretched human arms. Sometimes sharks become so conditioned to the arrival of dive boats that just the sound of an engine revving up is the equivalent of a dinner bell. Video evidence of the practice with a diver feeding fish to an Oceanic Whitetip in the Red Sea was shown to the investigating team. 'They hold the fish in their hand and as the shark gets very close, they would release the fish and the shark would chomp down and swim behind them. At that point the diver would reach out behind and pull out another fish from his pack. Over a given length of time, sharks become habituated to it – just as your family pet would, if you were teaching it new tricks. That's why the animal has such determined behaviour around people. It is looking for something and if it doesn't receive it, there is a high probability that you will be bitten,' explains Ralph Collier of the Shark Research Committee.

Indeed the practice could have been going on in Sharm El Sheikh's waters for over a year and Collier believes that a shark trained to take fish from a diver may have approached some of its victims, thinking they had food. 'The first reaction a human has, of course, is to fend off and at that point the victims stretch out a hand and the shark bits it off, thinking it was a fish,' adds Collier. Red Sea rescue diver Hossam El-Hamalawy agrees: 'This should be a reminder that the ocean is the shark's natural habitat and that we are visitors there. When we begin messing with the inhabitants' behavioural patterns, when we begin messing with their environment, then the consequences can be serious. I'm surprised the government has just woken up and discovered this overnight – the problem of the tourist industry damaging the ecology of the Red Sea has been going on for three decades and yet nothing has been done about it.'

This would go a long way to explaining the abnormal frenzied attacks by the sharks – ceaselessly biting at humans rather than the usual one investigative bite before swimming off in search of more suitable prey. Three of the five Red Sea victims were bitten over and over. 'Most of the time a shark bites a human once to figure out what it could be. To bite multiple times is very, very rare but we realised that all the bites seemed to be in the same area of the body – hands, legs and buttocks,' says Collier. 'It is not unusual for a limb to be bitten by a shark because generally that is the easiest thing for the animal to grab when it comes up to investigate. However, it is highly unusual for the victims to sustain wounds both to the hands and the buttocks area by the same shark.'

The most shocking aspect of the shark-feeding video was that the creature involved was the killer female Oceanic Whitetip and the explanation would go some way towards explaining its highly unusual, continuous attack on Renate Sieffert. Biologist Dr. Bojanowski adds: 'None of the bites looked like a test bite checking for something to eat; it looked more like the female shark was feeding and had somehow crossed the line of not identifying people as a food source.'

Simon Rogerson agrees with this view: 'I've heard reports of these approaches getting more and more aggressive. I think the thing is that the longer you spend in the water with this particular shark, the bolder it gets, and the more it tries to test you. To most sharks, human beings aren't edible and I've no idea whether an Oceanic Whitetip could thrive off a human but these sharks have got a history of being a danger to people who are in the water for a long, long time.' Rogerson believes scuba divers are at less risk than snorkellers, though: 'Divers have more control, and more awareness, of what's around them. Also, in nature, if something's floating on the surface, it looks as if it's injured so it becomes more of a target. These attacks have been happening to snorkellers who are just pootling off resort beaches and the chances are they're not wearing wetsuits. There's all that white flesh and that seems to attract sharks – it's the colour of fish flesh, after all. Most sharks like to sneak up on their prey – they don't like being seen – and that's easier if someone's snorkelling.' Of the attack on Renate Sieffert, shark behaviourist Erich Ritter adds: 'In a way it was if the shark had an agenda. It was not exploration, it was not a defensive wound – it just really went after her.'

Another theory came from the CDWS's Hisham Gabr. He blamed Australian live sheep exporters, passing through on their way to the Suez Canal and throwing any dead sheep into the sea. 'I know for a fact that sheep have been thrown into the water by a boat,' said Gabr. 'I don't know the quantities, I don't know the numbers and I know it was more than once because divers saw it. They saw the sheep thrown into the water and there was a ship passing by, carrying sheep in the Gulf of Aqaba, passing through the Straits of Tiran.' Dr. Bojanowski adds: 'Oceanic Whitetip sharks have a very widespread interest in different food items and carcasses of different kinds. If carcasses are being dumped and are drifting to the shore this will have a great effect on the movement and distribution of the shark, which is a known scavenger.'

This was all disputed by Australian meat and livestock export manager Peter Dundon. He said that although dead sheep were thrown overboard, maritime law forbids this in the narrow waterway near Sharm El Sheikh, and added: 'To my knowledge, there's been no Australian livestock vessel through there in that period. There's restriction of 100km [62 miles] for the closest land that whole dead sheep are not able to be disposed of. My experience on the vessels is that the master, who is responsible for that happening, enforces adherence to those international maritime laws very strictly.'

Over-fishing may have forced sharks closer to shore to find food. Warnings of this type of danger had already been aired as recently as August 2010 – just three months before the shark attacks began. The use of large nets has decimated the sea's fish populations, as well as endangering coral reefs and other marine life. Mahmoud Hanafy, Red Sea governor, professor of marine biology at the University of Suez and environmental adviser to the Hurghada Environment Protection and Conservation Association (HEPCA), is a leading figure in the process of halting the devastating effects of over-fishing. He said: 'Sharks, turtles and marine mammals, including dolphins and dugongs, all fall prey to these acts of fishing.' The practice also drives sea creatures such as sharks – attracted by the abundance of fish lured by regular feeding from holidaymakers – into shallow waters. Hanafy also points out how the average 20,000 tons of fish caught every year in the Red Sea is 'far surpassing the recommended sustainable limit of between 900 and 1,500 tons.'

According to statistics published by the General Authority for Fisheries Resources Development (GAFRD) – an organisation affiliated to the Ministry of Agriculture, Egypt's annual fish catch is over one million tons, coming from the local waters of the Mediterranean and Red Sea, the Nile River and lakes, and fish farms but it's a challenge to bring home the realities of over-fishing to the 200,000 local fishermen and workers employed in the industry, whose livelihoods depend on it. They already know that they are catching fewer fish than during the 1990s. Efforts to combat over-fishing in the Red Sea include the Hurghada Declaration. Signed in June 2009, the Declaration seeks to ban all net-fishing and trawling in the Red Sea with the exception of the area north of the Gulf of Suez. It also aims to establish 'no-take' zones, making certain areas free of any fishing activity. The Hurghada Declaration – named after one of the Red Sea's other major tourist areas – was drawn up and signed by HEPCA in conjunction with the Red Sea Governorate, the South Sinai Governate and the Ministries of Agriculture and Environment.

Some theories were less convincing, with one being that the Israeli intelligence agency Mossad was behind the attacks in an attempt to wreck the Egyptian tourist industry. The governor of South Sinai, General Abdel-Fadeel Shosha announced this theory at a meeting (and no doubt went on to bitterly regret it). As witty observers noted: 'Whether this was an Israeli agent in a shark costume, a specially indoctrinated Zionist shark, or a remote-controlled cybershark, the General does not elaborate, but he does say the theory needs investigating.' Speaking on the public TV programme *Egypt Today* on 5 December 2010, a specialist introduced as 'Captain Mustafa Ismail, a famous diver in Sharm El Sheikh' said that the sharks involved in the attacks are ocean sharks and do not live in Egypt's waters. Asked how the shark entered Sharm El Sheikh waters, he replied: 'No, it's who let them in.' Urged to elaborate, Ismail said that he recently received a call from an Israeli diver in Eilat telling him that they had captured a small shark with a Global Positioning System (GPS) planted on its back, implying the sharks were monitored to attack in Egypt's waters only. On the Sky News Middle East blog Dominic Waghorn retorted: 'Israelis get blamed for a lot in this part of the world, but Egyptian officials have plumbed new depths of pottiness with their latest Zionist conspiracy theory.' Meanwhile, Israeli officials rejected the notion as 'ludicrous', with Israeli foreign ministry spokesman Igal Palmor telling the BBC: 'The General must have seen *Jaws* one time too many and is confusing fact and fiction.'

On 9 December 2011, Egypt's ministry of tourism confirmed it would offer $50,000 (£31,000) in compensation to each of the Russian tourists attacked by sharks. The money came from private tourist companies in Sharm El Sheikh. A spokesman confirmed: 'The payment will be made by private Egyptian companies, such as hotels and diving clubs, not the Ministry of Tourism.'

On 11 December 2010, the committee of experts, which as well as the Americans included Moustafa Fouda of the Ministry of State for Environmental Affairs, Mohammad

Salem, Egyptian Environmental Affairs Agency, and Nassar Galal, Chamber of Diving and Water Sports, presented its findings, which still did not contain the definitive reason for the attacks. The report summarised several theories:

- The illegal dumping of sheep carcasses by animal transport vessels within 1.9km (1.2 miles) of the shore.
- The unique underwater topography of the area – i.e. deep water very close to shore allowing pelagic sharks and humans to swim in close proximity.
- Although fishing is restricted in the Sharm El Sheikh region, uncontrolled fishing in the Red Sea has depleted fish stocks and reduced the amount of natural prey available to sharks.
- Shark and human population dynamics; five million people visit Sharm El Sheikh annually and numbers of sharks migrate through the area each year.
- Feeding of fish by glass-bottomed boats and swimmers drew the sharks close to the beach.
- Elevated sea temperatures resulted in higher metabolic rates in the sharks and increased their energy (food) requirements.
- Although prohibited, it is believed that some dive operators have been feeding the sharks, which could have habituated the sharks to humans as a source of food.

Nevertheless, the report recommended that the beaches at Sharm El Sheikh be re-opened. There was a list of conditions, including the erection of 6m- (19ft) high watchtowers manned by trained lifeguards at regular intervals, swimming being restricted to designated areas – and a total ban on 'recreational shark feeding'. Dr. George Burgess, who headed up the investigation, said that his team believed someone 'accustomed the sharks to being fed and whoever did it has stopped,' but he added: 'The sea is a wilderness, just like a jungle. It can never be made entirely safe for humans, in Sharm El Sheikh or any other resort – we can only try to reduce the odds.'

That same day, Olga Martsinko's daughter Elena spoke of how ill her mother still was. What was left of her arm had had to be amputated and she had undergone emergency plastic surgery to her thigh and buttock. Her body was swathed in bandages from where surgeons had taken skin from her hip to try and cover the hole on the buttock. Another operation was necessary. Said Elena: 'My mother can only lie on her tummy because there is a hole instead of a left buttock. Inside the hole you can see the base of her spinal cord. It is impossible for her to move around with such a large hole in her body. I told her we should go back to Moscow to a hospital there. She said she would try to stand and walk a bit, but she lifted herself up and fell back, unconscious from the pain.' Six weeks on, Olga was still in hospital.

A week after Elena's statement came the official announcement: 'Tourism in South Sinai has absolutely not been affected'. Indeed, the Red Sea resort of Sharm El Sheikh had actually seen a rise in tourists booking for the Christmas and New Year season. The authorities introduced $50,000 (£31,000) fines for hotels and diving centres feeding the fish and fines of up to $15,000 (£9,000) for tourists who also did so. In addition, there are penalties for diving boats throwing waste into the sea or allowing clients to feed fish. These include suspension of the diving business from one month to six months and a withdrawal of their licence in the case of reoccurring violations.

Not everyone shared the optimism that all this would bring back tourism to the area, however. Hamdi Abdelazim, economy expert and former president of the Cairo-based Sadat Academy for Administrative Sciences, perhaps shared the thoughts of many: 'It is only natural that tourism, especially resort tourism, would be impacted following a string of shark attacks.' And, as one tour guide noted: 'Tourists come just for the sea, so if there is no opportunity to go to the sea, there is no reason to come here.'

In fact, the massive numbers of tourists entering the Red Sea every year could even be part of the ever-growing shark problems. 'Population dynamics are one of the primary factors in shark-human interactions. The fewer the people, the less likely chance you are going to have of encountering a shark. The more people you have, you increase that probability. Now you have interactions between the animal that is hunting and humans, especially when you look at the number of people who utilise the resort over the year,' said Ralph Collier of the Shark Research Committee. Until the 1970s, Sharm El Sheikh was a remote Bedouin fishing village with an empty, unspoiled coastline. Now it is the most popular resort in Egypt, with just a 8km (5-mile) stretch of beach to accommodate the millions of swimmers, snorkellers and divers. Hotels, keen to cash in on the allure of the water, coral reef and prolific sea life, all have their own jetties and pontoons from the beach.

On 14 December, the CDWS issued an update announcing that although beaches were open, there were still restrictions on diving and watersports in the Red Sea 'while safety assessments continue following the shark attacks.' The Chamber said that it wanted 'to reassure its members that it is constantly monitoring the situation and gradually hopes to lift the restrictions in the near future. However, the organization has underlined its priority in any decisions in the safety of visitors and its members.' Any divers still wishing to enter the Red Sea now had to be fully qualified with at least 50 logged dives and could only enter the waters with CDWS members in the area of Tiran, dive sites south of Naama Bay to Ras Mohammed National Park and the park itself. Diving was still completely banned between Ras Nasranie to the north of the Naama Bay jetty and no shore diving permitted anywhere in the Sharm El Sheikh area, with the warning 'Under NO circumstances are introductory or training dives permitted to take place in the sea anywhere in Sharm El Sheikh until CDWS members are advised otherwise'. Snorkelling and other watersports were still restricted along the whole of the Sharm El Sheikh coastal area, with the exception of glass-bottomed boat trips.

The South Sinai Hotels Chamber held a meeting attended by tour operators, tourist officials and South Sinai governorate officials in an attempt to explain the confusion over just how safe the Red Sea was. South Sinai Governor Major General Mohamed Abdel Fadeel Shousha said: 'This situation will continue until all technical studies, topographic surveys of the area and surveys of the sea bottom are executed by the researchers of the Suez Canal Authority. These experts from several fields are examining ways to secure the area in order that snorkelling can resume. All commercial vessels that enter or pass by Sharm El Sheikh will have to be accompanied by Egyptian officials until they exit Egyptian waters so as to make sure they don't throw dead animal or harmful substances into the Red Sea.'

General rules for safety when divers encountered sharks were issued by Dr. Elke Bojanowski:

- Only enter the water if you are comfortable with the situation and confident that you can stay calm.
- Do not enter the water if there is any sign of feeding activity around the boat.
- Be aware that you are most vulnerable on the surface, so control your buoyancy at all times.
- Avoid erratic movements.
- If you want (or need) to leave the water, do so in a calm and orderly fashion.
- Try to avoid surfacing straight above a shark swimming below you.
- To avoid Oceanic Whitetips coming too close for your comfort, staying next to, or retreating to the reef might help.
- Do not try to touch or in any way harass a shark.
- Do not be alarmed by a shark calmly circling you. Just make sure to turn with it and keep it in sight.
- Stay alert and look around you from time to time to see if another shark is approaching you from behind/underneath/above, otherwise one might sneak up to you.
- Generally, sharks are more reluctant to closely approach groups of divers than single ones.
- Remember, you are in the water with a wild predator whose behavior will never be 100% predictable!

On 22 December 2010, the saga of the serial shark attacks took a strange turn with the claim that the Oceanic Whitetip responsible for the attacks had been killed by a drunken Serb. This bizarre story involved a man called Dragan Stevic allegedly hurling himself off a diving board into the Red Sea, landing on the head of the shark and instantly killing it. The report even included an account from Stevie's 'friend' Milovan Ubirapa, who witnessed the incident: 'Dragan climbed on the jumping board, told me to hold his beer and simply ran to jump. There was no time for me to react or try to stop him; he just went for it. He jumped high and plunged down to the sea, but didn't make as much of a splash as we thought he would.'

This hoax account became an internet sensation but was dismissed when its source was revealed as a satirical American newspaper called *The Onion*. The shark featured was a Basking shark – and one that had washed up not on the beach at Sharm El Sheikh at the height of the terror reign but off the coast of North Carolina, a year earlier.

The threat of further shark attacks diminished when they migrated away from Sharm El Sheikh at the beginning of 2011, but they return each autumn. And the female serial-killer shark is still alive. It is highly likely that when she returns to the crystal-clear waters, she will again turn them red. Said Ralph Collier: 'As much as I am against euphemising any animal, I believe this shark has become habituated to humans and should be removed from the environment because its potential to do this again is very high.' Dr. George Burgess agreed: 'To catch that animal, you are going to have to find it first – that's a lot of expenditure in human time. But in the end, what have you got? Sure, you have some retribution for what it did, but you have no assurance it won't do it again, and no assurance its mates won't do it again. These are open-ocean sharks that are living in an environment that is food-poor so when you do find food, you darn well better take advantage of it! Do they remember things? Sure, they remember where the good places to eat were and they'll come back but they are not connect-the-dots sort of animals. They are basically swimming, sensory machines. Sometimes we make mistakes and sometimes they make mistakes. And sometimes we just happen to be in the wrong place at the right time – for them.'

Sharm El Sheikh resident 'Crowley' reflected on the events in the Red Sea from a professional diver's viewpoint in *The Equalizer* online magazine:

> It's been an interesting few weeks. The restrictions on training or intro dives and snorkelling have meant a downturn in business for a lot of operators, and media hype and misinformation have not helped – some people have literally cancelled their diving experiences because they were afraid to get into the water. Others were understandably unwilling to pay extra in order to relocate to Dahab every day. For the staff it's been uncertain. You can't work if you can't get in the water and most dive professionals in Sharm are paid in commission so for sure the shark attacks have impacted our income this month. Having said that, the impending collapse of the Eurozone and frozen airports have not been of assistance.
>
> Talk of sharks has of course been the buzz of the town with everybody turning into armchair Sharm El Sheikh

shark experts overnight. Our dive briefings included some extra pointers on how to deal with a shark in close proximity, and I think many guides – including myself – would admit to a few nerves at certain dive sites, particularly since the female thought to be responsible for some of the attacks (including the fatality) has been seen at Shark and Yolanda quite regularly.

The 'likely causes' for the attack were already the centre of discussion around the beer table long before any real scientists arrived. This is not meant as a slight to the experts – but it would seem there is a wealth of information here that nobody ever used. With a few notable exceptions such as Dr. Bojanowski, there is very little research into shark populations in the region and yet for thirty years this has been one of the most popular dive destinations on the planet. We write shark encounters on the 'Daily Sightings' board every few days in summer – and most of that doesn't get any further.

I don't think the dive industry will suffer long term but the last few weeks have made some people think (and a few unemployed!) and a tragic event might have a long-term positive outcome if more research and education into sharks in the Red Sea is properly implemented. We are all very hopeful that the situation will improve for the tourists, for the staff here – and indeed for the shark population. Until then, it's still beautiful here!

For sure, the last chapter of this tragic series of events has not been written and there have been rumours of sharks being caught and killed. The question is, if this has been as retribution or as a preventive measure, or if there has been a reason at all.

According to Ahram Online the South Sinai Governor's secretary, General Ahmad Saleh, said in an interview that there are watchtowers with professional lifeguards checking movements by sharks in the water. In addition, Zodiacs are patrolling the area looking for signs of danger. Furthermore the authorities are searching for the best places to install protective nets.

One question is what the lifeguards in the towers are looking for exactly? It's hard enough to spot a shark while diving, so how can it help to look out from the beach? Another question is what good the nets will do and how they have to be designed to fit the underwater topography of the Red Sea. Historically, those kinds of nets have caught more sharks going out to sea than coming into the beaches, plus managed to kill turtles and other marine life.

All diving restrictions put in place following the shark incidents in Sharm El Sheikh were lifted 22 December 2010. *The Equalizer* extends our sympathy to the victims and their families.'

On 13 January, the Egyptian Government of the South Sinai banned all fishing in Sharm El Sheikh, as well as the resorts of Dahab, Nuweiba, Taba and all along the Gulf of Aqaba. The move followed a series of findings that the shark attacks may in some part have been linked to the depletion of their natural food source and overfishing.

On 25 January 2011, the *Sharm Business & Community Magazine* aired its discontent over what it saw as the continued lack of effort by local authorities to ensure the Red Sea resorts were luring back visitors:

> We understand that some people won't be happy to open the subject of shark attacks again, hoping instead that it will be forgotten. The magazine would like to tell them not to put their head in the sand like an ostrich and stop pretending that nothing has happened. It is a fact that there has been an impact on tourism and that impact has led to a bad economy in the city. We know that the relevant authorities who are supposed to be in harmony to solve the dilemma are not doing enough. The situation, according to the information we have, is still not clear. We have heard a lot of announcements in the local and world media from unprofessionals. These announcements have served only to make everybody more confused. The magazine has received many mails and calls from readers and tourists from different nationalities asking whether they can come to Sharm and use the beaches yet or not. We have been trying to reassure them that what happened was an unusual series of accidents, and that the city is full of other alternatives which they can enjoy so much.

In February 2011, reporter Jennifer Reade noted that the Red Sea resorts had started to see 'a small increase' in tourism, before adding: 'After all the shark attacks that took place during last year, most of the beaches were left empty and some souvenir sellers and other retail establishments who have businesses in Sharm El Sheikh say they are struggling to stay afloat. A couple of sun seekers dressed in their summer gear explored the predominantly desolate town centre with its abandoned Bedouin-style coffee shops. One of the tourists from Devon in south-western England, who was travelling with his wife and ten-year-old twins said he is pleased that the town is empty and that there are no Russians around. The family booked their holiday prior to the shark attacks.'

That same month *travelweekly* reported that travel companies were continuing to fly holidaymakers out to Sharm El Sheikh, but this announcement had nothing to do with shark attacks. Massive anti-government demonstrations were rocking Egypt's main cities and reassurance was needed that they were a long way from the Red Sea tourist hot spot, where the tourist trade was already so badly damaged.

On 23 February the South Sinai Governor issued a 'decree' allowing all shore diving activities and shore house reef diving activities in all Sharm El Sheikh areas. Then, on 9 March 2011, a British couple told a chilling story to the *Sun* newspaper. Richard King, 32, and his partner Laura Hooper, 29, said they had watched in horror from a dive boat as a 5m (16ft) Tiger shark ripped at a female corpse floating in the waters off Sharm El Sheikh. They were less than a third of a mile from shore. Incredibly, the dive crew insisted on continuing with the trip and refused to act on what had been seen. Instead they told Richard and Laura (and five other tourists with them) not to mention the incident. Said salesman Richard, from Swindon, Wiltshire: 'The crew saw the shark soon after leaving Sharm Beach for a day's scuba-diving. As we got closer, we clocked that it had been feeding on a human body. Laura ran into the boat in tears. We were asked by the crew not to say anything as it would be bad for tourism. We couldn't believe it when there were people swimming in the water the next day – there was no mention of sharks in the water or a dead body.'

Indeed so horrified were the couple by the whole event, they cut short their holiday and flew home. The dive centre concerned insisted the crew had reported the body during the trip and that all the holidaymakers had said they wanted to continue. Police later recovered the body and began an inquiry. A spokesman for the tour operator insisted the incident had been reported by the dive crew on their return but added: 'It was not a shark attack that killed that person.' Dr. Adel Taher, director of the Hyperbaric Medical Centre in Sharm El Sheikh, said he examined the woman's body and that her death was not due to a shark attack but due to drowning and subsequent propeller injuries.

The story of the Red Sea sharks goes further: not only are they the hunters but the hunted, too. In June 2010, six Yemeni fishing boats were intercepted and found to have several lengths of long lines on board, as well as more than twenty tons of dead sharks. This was despite a 2005 ban on shark fishing in a bid to halt declining numbers of the world's once-highest shark population. Most local shark fishermen had heeded the ban but during the shark season of 2010, an influx of foreign fishermen once more threatened the sharks of the Red Sea. Said Amr Ali, HEPCA's managing director: 'Our local fishermen were never interested in sharks – nobody likes the meat; it's bulky and sells very cheaply. About a decade ago the Chinese started showing up here. They taught our fishermen that they could get good money by selling the fins and they didn't have to bring the whole animal back. They could just cut off the fins and put the rest back in the water.'

Egypt's ban is only effective within 20km (12 miles) of the coastline. Beyond that, the waters of the Red Sea are fair game for fishermen. Dive boat captains have reported that off the coasts of Sudan and Entrea, long-line shark fishing – long fishing lines with thousands of baited hooks – is 'out of control'. In an attempt to curb shark fishing, the Indian Ocean Tuna Commission (IOTC), in charge of managing fisheries in the Red Sea, has implemented regulations. These state that shark fins must not exceed 5 per cent of the weight of the shark carcasses on board, but this is hard to control. Said Ali: 'Our shark protection laws have given sharks in the Egyptian Red Sea a chance to reproduce safely, but it has also made us a target for every shark fisherman in the region.' As one naturalist observed: 'It would be a shame to see the Red Sea sharks suffer a similar fate as those of the Mediterranean. However, as we are seeing now, the fishermen have apparently moved into the precious Red Sea at full steam. Unless action is taken immediately, we may ultimately witness the decimation of one of the world's last remaining shark strongholds.'

In July 2011, HEPCA announced an all-out programme to preserve the sharks of the Red Sea:

> Shark populations worldwide have been pushed to the brink of extinction within the last few decades. Millions die every year because of the insatiable human demand for shark fins and the highly destructive, unselective fishing methods practised by the industry around the globe.
>
> The sharks' disappearance is worrying on a number of levels. As top oceanic predators, they are of fundamental importance to the balance of the marine ecosystem. Removing them on a large scale has severe consequences through succeeding layers of the marine food web. It has

altered other species' abundance, distribution and diversity, and impacted the health of a variety of marine habitats, including sea grass beds and coral reefs. A recovery from depletion is hard to accomplish since most of the larger shark species have a very low reproductive potential; they take years to reach sexual maturity and produce very few young.

Besides ecological considerations, the high economic value of living sharks is being recognized by more and more countries. Shark diving is a growing industry; the interest in encountering sharks underwater creates far more revenue than the one-time profit to be gained from killing them.

All these factors have researchers and conservationist campaigning to convince lawmakers and governmental bodies to finally step up and include more shark species in their protection schemes. Their efforts are hampered by the lack of available information on basic population parameters and life history patterns of many of the depleted shark species, which are necessary to create and implement effective conservation programs.

Similar problems apply to the numerous shark species living in the Red Sea. Fishing pressure has (and has had) an impact of unknown magnitude, and – despite local or regional regulations – no protective legislation for the area as a whole exists. And while the sharks' economic value, especially for the Egyptian tourism sector, is undisputed, scientific data is scarce; information on population status and ecological needs are insufficient or don't exist at all.

To rectify that situation, the Hurghada Environmental

Protection and Conservation Association is launching a comprehensive research project on Red Sea Sharks, using such diverse methods as a volunteer monitoring program, photo-identification, electronic tracking, and dedicated surveys.

Overall aim is to collect data on species distribution, residency and migration patterns to identify critical habitats for feeding, mating and giving birth. It is this kind of information that will be crucial for designing and managing effective protection measures.

The diving professionals in the Red Sea will be called upon in the future to help in monitoring the shark populations in the Egyptian Red Sea. Their experience and ongoing daily activities in the water make them perfect volunteers, who can provide us with a wealth of information.

One species has already proven to be highly accessible to photo-identification techniques. Over the past 6 years, more than 20,000 underwater photographs of Oceanic Whitetip Sharks (*Carcharhinus longimanus*) have been analysed to create a catalogue that contains more than 500 individuals.

Greatly advanced tracking devices are being used to monitor the secret life of the shark. Tags that store information on temperature, pressure and light levels can be attached to marine creatures for longer periods of time. They release themselves from the shark at a predetermined time and then transmit the data to satellites passing above. In 2010 a pilot project was launched to fit two Oceanic Whitetip sharks with tags for one and two months respectively during

Egypt's winter season. Said Dr. Elke Bojanowski, who headed up the project: 'The data gained from these two individuals will provide the first insights into the sharks' diving behaviour and movement patterns in the Red Sea. These results will help us to design a more extensive future tagging project, ultimately aiming at identifying critical habitat for the species, including major feeding areas, breeding and nursery grounds.'

THE OCEANIC WHITETIP SHARK

This shark was nicknamed 'Lord of the Long Hands' by famous underwater explorer Jacques Cousteau. Despite being a somewhat sluggish creature, the Oceanic Whitetip is a clever hunter with its pale fins looking like a moving school of fish to its prey, allowing the shark to get close and launch surprise attacks. As we have seen, this shark is not afraid to investigate humans, too.

Description: White-ish tipped fins, sometimes accompanied by white mottling, with the dorsal (back) fin varying in colour depending on its habitat, e.g. bronze to brown in the Red Sea, greyer in the Indian Ocean and the pale beige of Hawaii. White underside.
Has broad, triangular, serrated upper teeth with overlapping bases and narrow-cusped lower teeth with serrated tips.
Adult male: 1.72–2m (5.6–6.5ft)
Adult female: 1.8–2m (6–6.5ft)
These sharks can grow to a maximum of 4m (13ft).
Diet: Squid, schools of fish, stingrays, seabirds, turtles, dead marine animals (including whale carcasses) and general ocean animal debris.

Habitat: Open ocean, coral and rocky reefs, preferring depths of at least 150m (492ft).
Found in the following waters: Central Pacific, Tropical Eastern Pacific, North and South Atlantic, Western North Atlantic, Caribbean, Amazonian, Chilean, Argentinean, Eastern North Atlantic/Mediterranean, West African, Southern African, Central South Indian, Madagascan, Arabian, Indian, Southeast Asian, Western Australia, Southeast Australian/New Zealand, Northern Australian, Japanese.

THE MAKO SHARK

The Mako is one of the fastest sharks in the ocean, reaching speeds of up to 96.6kph (60mph), and can leap as high as 6m (20ft). It is famous for its amazing displays of power and strength when chasing prey. The Mako doesn't eat large animals or humans but it doesn't like humans in its environment, viewing them as a threat.

Description: There are two kinds of Mako; the more common Short Fin, and the Long Fin. The Short Fin is a sleek, spindle-shaped shark with a long conical shout. It has short pectoral fins and a crescent-shaped tail fin. There is marked countershading on this shark; its dorsal fin a metallic indigo blue, the underbelly white. There is a distinct keel on the caudal (tail) base. Its second dorsal fin is much smaller than the first. The long fin is slimmer and has broader, straight pectoral fins and a moderately long, conical and pointed snout; the pectoral fin is as long as its head (and sometimes longer). Its eyes are large and it has a parabolic shape. Colouring is dark blue or grey-black, with a white belly and dark jaw.

Short Fin: Has slender and slightly curved teeth with no lateral cusps, which are visible even when the mouth is closed.
Long Fin: Large and blade-shaped, with no lateral cusps or serrations. The lower anterior teeth protrude from the jaws and are in line with the shark's lateral teeth.
Size: Can grow to a maximum of 4m (13ft).
Diet: Large fish, including tuna, bluefish and swordfish, but smaller species too, such as mackerel, herring, cod, Australian salmon and sea bass. May also feast on other sharks, porpoises and turtles.
Habitat: Tropical and temperate offshore waters, preferring to dwell as deep as 150m (492ft).
Short Fin: Found in the Western Atlantic, from Argentina and the Gulf of Mexico to off the coast of Nova Scotia, Canada, and Eastern Atlantic, Mediterranean, the Indo-West Pacific, Central Pacific and Eastern Pacific. Rarely seen where the water temperature is less than 16°C (60.8°F).
Long Fin: Sighted in the Western and Eastern Atlantic, Western Indian Ocean, Western Pacific, Central Pacific and Eastern Pacific, with reported, unconfirmed sightings in the Mediterranean.

DWELLERS OF THE DEEP

The Oceanic Whitetip and Mako sharks that terrorised swimmers and snorkellers are just two of the 30 or so different shark species in the Red Sea. Their presence so close to shore was certainly unique for most of sea's sharks are rarely seen, although the Oceanic Whitetip is the most likely to be spotted, some preferring to circle waters so deep that even divers will never come across them – hence the investigation into the extreme circumstances which drew the

Oceanic Whitetip and Mako into shallow waters and contact with humans. Listed below are just some of the Red Sea sharks:

- Grey Reef shark
- Scalloped Hammerhead
- Blacktip
- Dusky shark
- Spinner shark
- Silky shark
- Tawny Nurse shark
- Pelagic Thresher
- Grey Nurse shark
- Great White
- Sandbar shark
- Tiger shark
- Oceanic Whitetip
- Shortfin Mako
- Blue shark
- Whale shark
- Great Hammerhead
- Spiny Dogfish
- Whitetip Reef shark
- Bramble shark
- Zebra shark
- Coral Catshark
- Whale shark
- Hooktooth shark
- Sliteye shark
- Spottail shark
- Snaggletooth shark

- Bull shark
- Bignose shark
- Bigeye Houndshark
- Arabian Smoothhound
- Sicklefin Weasel.

CHAPTER 2

HONEYMOON HORROR IN PARADISE

'I HEARD "HELP!" AND THE MOST AWFUL SCREAM, AND I CAN STILL HEAR IT WHEN I CLOSE MY EYES...'

One blissful moment Gemma Houghton was an excited newlywed, the next she was a traumatised widow – and witness to her husband's terrifying death by shark.

The setting had been idyllic, the weather glorious and the waters a dazzling blue. Then a shark came into view and fatally savaged British software creator Ian Redmond. The couple's honeymoon beside the Indian Ocean came to a sickening end.

Gemma, 27, had been sunbathing when she heard cries from the ocean at Anse Lazio beach on island of Praslin in the Seychelles. Her husband of just ten days was caught in the jaws of a shark as he snorkelled in the waters; within minutes he was dead. Recounted one onlooker: 'I saw the swimmer who was missing a huge chunk of flesh from his left leg, so much so that I could see the bone of his thigh. He was sickeningly pale, but still had his flippers on both feet. At this point, a woman ran over and started screaming "That's my husband!"' Others on the crowded beach had looked on in horror as 30-year-old Mr Redmond's cries of 'Help! Help!' reached their ears. The hand of his left arm, ripped off in the attack, still bore his wedding ring.

The attack happened around 4.30pm on 16 August 2011. Mr Redmond had been snorkelling just 10m (33ft) from shore. After marrying in Britain's Yorkshire town of Skelmersdale, the couple were in the second week of their honeymoon. They had chosen a beach described as one of the most beautiful in the world but this beauty had become an ugly reality of sharks' existence – for not only had Ian become a victim, but two weeks before so too had another holidaymaker.

Despite rescuers in two boats rushing to his aid, Ian was pronounced dead from massive blood loss upon arrival by helicopter at the Victoria Hospital. Efforts by a French doctor to save him at the beach had failed: Redmond had lost an arm and flesh from his leg, and also suffered wounds to his chest and stomach. Those close by at the time of the attack described how they had seen the shark in the water. It was still there, its fin clearly visible, when the victim was pulled into the dinghy. 'Someone had seen a fin sticking out of the water and then we saw a dinghy pulling a man from the water,' said an American tourist. 'Someone grabbed the woman and tried to keep her away. People all over the beach were just hugging whoever was close to them or trying to keep any children from witnessing what was going on. The damage was too great. We kept his wife away from the body because it was too horrific for her to see.'

Jeanne Varigolu, owner of a beach restaurant, said he had overheard Gemma saying she 'still had hope' for her husband as he lay horrifically injured on the sand. Chantal Andre, a restaurant owner, accompanied Gemma to the hospital – 'It was just awful. She wasn't crying at all, she just couldn't believe he was dead,' she said.

Gemma would later speak of those last moments: 'He has never screamed like that before because he is such a strong man, so brave. All of a sudden I heard "Help!" I thought at first he was sneezing. Then I heard it again. I heard "Help!" and the most awful scream, and I can still hear it when I close my eyes. He looked up at me and I looked up at him, and I could see a mixture in his eyes of fear and of realisation – relief that he had seen me and that I was there. I reached out my hand and held his face, and I got his hand and held it to my chest and I said to him, "You are going to be all right – we're going to look after you. We are going to sort you out." I think I told him I love him very much – I hope I did.'

Tragically, she recalled how her husband had teased her when she asked a hotel receptionist if there was any threat of sharks and was told there wasn't. 'Ian had laughed at me when we were on Denis Island. I overheard a man teasing his wife, saying there were sharks and things, and I asked a lady on reception and she said, "No, not in the Seychelles – they are very safe waters."' Gemma told the BBC: 'One of the reasons we picked to come to the Seychelles was the beautiful waters, the fact that it's like an underwater aquarium and there's not really any dangerous animals. I've heard of stone fish and you have to be aware of the currents when you're snorkelling, but we didn't think that sharks would be in the Seychelles at all.' This, however, was disputed by those Gemma said she had questioned. Richelieu Verlaque, owner of the Bonbon Plume restaurant, where Ian and Gemma had enjoyed a seafood lunch shortly before he died, said his partner, Chantal Andre, had noticed Mr Redmond's snorkelling equipment and warned them that a shark had killed a French tourist, just two weeks earlier.

'The couple were having lunch here. Chantal was saying, "How was the lunch?" The woman said "Paradise." They were here because her parents had come here a few years ago and recommended it. The warning was to be careful, there's been a shark attack, don't go beyond waist deep. I heard the conversation. She said, "Yes, we're aware – we've heard about this incident."' Colleagues at the restaurant said Ms Andre was distraught that Mr Redmond had ignored her warning. At the beach she had comforted Mrs Redmond and later accompanied her to the hospital, Mr Richelieu said. Police spokesman Jean Toussaint said Mrs Redmond's claims that she and her husband were unaware of the threat were 'unbelievable.' The death of the French tourist had been widely reported in the Seychelles, especially on Praslin, he said.

Toussaint confirmed that Redmond had lost an arm during the attack and sustained other serious injuries: 'He was assisted medically, but unfortunately he could not make it. He had no chance of surviving because of the nature of the injuries.' The Foreign Office officially confirmed Redmond's death and Gemma issued a heart-rending plea that her husband's wedding ring might somehow be retrieved. Said one tourist: 'They've been searching for the arm in the sea in the hope they will find the ring. They may yet find it inside the shark, if they hunt it down and kill it. It's hard to imagine a worse time for a bride: she's lost her husband, now getting back his wedding ring is all she has left.'

Meanwhile, a press release issued by the Seychelles authorities outlined a plan of action following the shark attacks:

1. The tooth of the shark will be sent to South Africa for further research to determine the identity of the shark. DNA tests are to be conducted on it.
2. Erect two exclusion nets at Anse Lazio as soon as possible to facilitate swimming in the area and another one at Petite Anse Kerlan. These will be set up as trials and as temporary measures. Praslin Development Fund has been given the responsibility to lead this particular activity.
3. In an effort to provide greater safety for swimmers the Seychelles Fisheries Authority has been given the responsibility to coordinate and manage the use of drumlines just outside these swimming areas.
4. Seychelles Fisheries Authority will intensify their research on bull and tiger sharks in Seychelles coastal waters. The study exists already but now will be given higher priority.
5. Seychelles Maritime Safety Administration is to intensify its work in enforcing against pleasure boats that dispose of their waste at sea, especially in coastal waters.
6. Seychelles Maritime Safety Administration will import emergency shark attack first-aid packs and place them at strategic points on popular beaches. The Administration will employ and train lifeguards in using them.
7. Monitoring and surveillance plans will be developed for strategic areas.

8. An emergency protocol will be developed and personnel of key agencies will be trained in using them.
9. The South African experts will develop a list of best practices that will be shared with all concerned.
10. The Seychelles Government will make a formal request to the KwaZulu-Natal Shark Board with the aim of establishing long-term cooperation between the Board and the Seychelles Government. The aim is to obtain the assistance of the South African experts to assist government in building a robust institutional framework to address the local shark problem.

Tour operators maintained that although the shark attacks were 'devastating', bookings to the Seychelles were still coming in and that the attacks would have only a 'short-term' effect on business. One commented: 'We had one couple who came in yesterday afternoon, who said they'd heard about the shark attack but it didn't bother them. A lot of our customers are repeat clients who know the area, who know how safe it is. But if anyone is concerned and wants to cancel a holiday, then of course we would honour that.'

But which beast from the deep was the predator this time? A tooth was indeed removed from Mr Redmond's body by doctors and sent to the KwaZulu-Natal Sharks Board in South Africa, where it was initially revealed the killer was a Great White, though spokesman Jeremy Cliff admitted he had a gruesome task to perform: examining Mr Redmond's body to confirm the shark's identity. 'We really need to examine the victim because it is better to see the actual wounds rather than photographs. We are not shark hunters, we are scientists and we want to help Seychelles authorities to protect their beaches,' he said. Cliff was right in his determination to track down the shark for it had already killed a fortnight earlier. Teacher Nicolas Francois Virolle (36), from Rodez, France, was attacked as he swam just 15.5m (51ft) from the shore on 1 August 2011, the last day of his holiday. He was heard to shout 'Shark!' before disappearing under the water. His savaged body was found by a fishing boat. Fellow tourists jumped into the sea to try and rescue Mr Virolle but he died from a massive loss of blood.

Local resident Dominique Pothin described the horror: 'We were on the beach in front of the Bonbon Plume restaurant on the corner. There were dozens of tourists on the beach. Around 3.30pm it was still beautiful, just as it had been in the morning, and the water was calm. If there had been a flag, it would have been green. The man was swimming in front of us, about 50m (164ft) from the beach. He started screaming and had some sort of nervous laughter bordering on hysteria. We thought he was an idiot. He ended up calling for help and he clearly said "Shark!" Then, for a very short time we did not see him – he was caught at the bottom. The water had become all red around him. He then reappeared. Two Seychellois in a fishing boat came to his rescue; the poor wretch just about had the strength to lift an arm. The two Seychellois managed to take him back on the beach. My wife, like many others, went to see the boat: the victim had almost no belly. I went to the restaurant and asked to call for help; someone else had already called an ambulance. The victim remained more than half an hour in this boat, both legs and feet dangling over the side. A lady, perhaps a doctor, finally got into the boat and there was an umbrella to protect the poor man from the sun. This gesture seemed ridiculous but at the same time full of humanity. A tourist at the back of the boat said there was nothing left to do – the swimmer was dead. People were standing on the beach, staring. Some cried, others stood motionless; I thought about my son, the fragility of things.'

After Ian Redmond's death, Nicolas Virolle's father Alaine, 65, launched a furious attack on the authorities. Grieving his son's 'violent and sudden death' on a 'lazy, safe beach holiday,' Mr Virolle also despaired that someone else had been allowed to fall victim to a shark. He accused those in charge of the resort of failing to alert holidaymakers for fear of losing vital tourist trade: 'It's an absolute tragedy and one which could have been avoided. It was criminal to leave the beach open in these circumstances – you can't run a tourist resort full of hundreds of holidaymakers and expose them to these kind of dangers. It appears that the authorities in the Seychelles are more interested in making money than ensuring people's safety; they are obsessed with money. It is a disgraceful situation. In the photos I've seen of him on the internet Mr Redmond looks very like my son. He was about the same size and build, about the same age and had the same kind of face – the thought of both of them being killed this way is almost too much to bear. Mr Redmond was married, of course, which perhaps makes it even worse for his family. It is terrible to think of the distress of his poor young wife – my family sends her all our condolences.'

The retired English teacher said that initially Seychelles officials even refused to accept that his son had been killed by a shark, claiming he had been involved in a collision with a propeller: 'I didn't believe this at all. My son was an excellent swimmer and had never got into difficulty in the water. He was in a group who were all looking out for each other and there were more than 100 people on the beach when the accident happened. All the Seychelles officials wanted to do was make out that Nicolas had got into trouble by himself and that the resort was entirely safe. They even sent me a card and a bunch of flowers to try to win me over. The Seychelles authorities should have launched an inquiry straightaway to work out why the shark was coming so close to the beach. Perhaps food was being thrown from boats? We now need to put pressure on the Seychelles Government to deal with this matter seriously. The situation cannot be allowed to continue: lives are more important than profits. We have lost Nicolas but there is no need for others to suffer. My family and I have our memories and are trying to immerse ourselves in our work to deal with our loss. It is an awful time for everybody involved in this. We have received condolences from all over the world, which has helped enormously, but must live with our grief.'

British holidaymakers, too, accused the local authorities of trying to keep the circumstances of the death of Nicolas Virolle quiet for fear of driving tourists away. Said one, 45-year-old David Simpson: 'It felt like there was a news blackout. There was nothing in the *Nation* newspaper and there was nothing on the TV: it just seems that it wasn't reported. There was no mention of it. We assume Mr Redmond would have known nothing about it. Surely that beach should have been closed. They should have put up warning signs, but I think they are putting people's safety at risk to protect the image of the place.'

Contributors to internet forums accused the Seychelles

authorities of culpability for Ian Redmond's death, likening officials' response to the plot of the *Jaws* movie, in which the mayor of a beach resort refuses to accept swimmers are in danger despite a series of shark attacks. Gervais Henrie of the opposition Seychelles National Party said the islands had a 'culture of silence' in which the state-controlled media was told not to run stories that might damage the all-important tourist industry. One holidaymaker, back in Britain after his visit to Praslin, spoke of his outrage at how the shark attacks had been played down:

> I still can't believe that this has been allowed to happen. The second shark attack could have been prevented if the authorities had acted more sensibly and learnt from other, similar incidents (Egypt recently). The issue is that as a tourist, everyone you speak to regarding your holiday (agents, hotels, locals) has a vested interest in promoting tourism. We checked into our hotel and immediately were marketed a yacht trip by our travel agent around Praslin to visit islands ending at Anse Lazio for snorkelling: the agent, a local, left out the fact that there had been a shark attack in the last week on the very same beach. On the boat, we asked the crew about sharks and they told us about the shark attack as if it was a minor incident, leaving out the fact that it was fatal! Then, when we arrived at Anse Lazio, the crew changed their minds about snorkelling there due to the "high tide" (!) and took us to La Reserve to snorkel, which is a few miles away (enough for a shark to swim to).
>
> We swam there in deep sea without seeing many

> fishes. The boat owner threw pieces of bread into the sea to attract the fish for snorkellers! Lots of people went snorkelling without knowing about the attack and even on Anse Lazio, there was no sign on the beach about the incident and people were swimming cluelessly. We only found out about the gravity of the first attack after we reached our hotel and Googled the news! After that, we didn't swim in the deep sea and checked the news every night as we were scared that there would be another attack (just like in Egypt). We arrived back in London yesterday and couldn't believe the news this morning. This could have been prevented in my opinion, if the authorities had taken measures to warn tourists.
>
> In general, we found the local boat operators very callous regarding safety, even while swimming – there are strong currents at this time of the year and lifejackets are not provided for snorkelling. Promoting tourism is one thing, disregarding safety and hushing up real dangers is criminal. I wouldn't recommend anyone (let alone with kids) to go back there right now.

But some, like Tim Ecott, author of *Neutral Buoyancy: Adventures in a Liquid World*, were of the view that it was 'unrealistic' to close Anse Lazio beach after Virolle's death. Writing in the *Telegraph* newspaper, he said: 'People in the Seychelles are already talking about offering rewards for the killing of large sharks. There is also talk of the shark responsible being a "visitor" rather than a local shark. As with all shark attacks, the incidents at Anse Lazio have resulted in wild speculation, imagination and fear gripping the island. Expert opinions are now being sought and details of an autopsy report from Ian Redmond's body will hopefully reveal the identity of the species responsible. For now the Seychelles Government is being proactive, sending vessels and fishermen from the National Parks Authority and the Seychelles Fishing Authority to Praslin. Beaches around Praslin have been closed and tour operators have suspended snorkel-based activities until further notice. Unfortunately, when it comes to sharks there is never an easy answer. When we enter their world, they have all the advantages and sometimes we become their prey.'

One local resident, Allen Hourareau, agreed: 'Sharks, animals and humans are the same. You start feeding them with free meals and you will see them at the same spot all the time. If there are sharks everywhere then why only one attack during those years? People feeding sharks in the Seychelles are mostly tourists and divers – we Seychellois, we fish and *eat* them, like you see on the marketplace daily. Please do not worry to swim in the Seychelles – it is the safest place in the world. *Please* let us not make this incident a big issue and scare the tourists away, as this site does: remember our economy depends on tourism. Thanks once again.'

Director of the Seychelles tourism board, Alain St Ange, said the attack on Virolle was by a 'foreign shark' and was therefore a 'freak accident'. Police officer Jean Toussaint agreed, but said 'a big effort' was being made to 'get this beast out of the waters.' Now an emergency meeting was held on 15 August, after which the Seychelles' interior minister Joel Morgan announced all nearby beaches would be closed and swimming was banned:

With immediate affect, the Seychelles Maritime Safety

Administration has issued a temporary ban on swimming or entering the water in certain bays of the island of Praslin and islands off the coast of Praslin following two fatal shark attacks this month in the northwest area of the island. People should not swim or enter the water in the following areas:

1. Anse Lazio – patrolled by the Police and National Parks Authority.
2. Anse Georgette, Petite Anse Kerlan and Grand Anse Kerlan – patrolled by Lemuria Resort.
3. St Pierre Island – patrolled by the National Parks Authority.
4. Curieuse Island – patrolled by the National Parks Authority.

This measure will be enforced for a certain period in order to search for the shark and remove it from the water.

For all other areas of Praslin and La Digue, people should swim close to the beach, within a short distance and not in deep water. The Seychelles National Parks Authority in collaboration with the Praslin Development Fund, Seychelles Coast Guard, the Police, the Department of Environment and the Seychelles Fishing Authority, as well as several private boat owners and Seychellois fishermen, will be undertaking continuous patrols, research and fishing activities in order to capture the shark.

The Maritime Safety Administration will also reinforce the ban on dumping of waste from yachts and other boats, which have been reported in some of these areas.

The last recorded fatal shark attack in Seychelles waters was in 1963 and therefore the two attacks are considered as extremely unusual events.

On the very same websites reporting the attacks was the somewhat hollow invitation: 'Don't let this attack put you off visiting the stunning Seychelles. Discover some of its best villas here' followed by a list of the 'top luxury villas in the Seychelles'. It was pointed out Britain's glamorous Royals, the Duke and Duchess of Cambridge, had honeymooned only two months before at North Island, just 19 miles (30km) away. Michel Gardette, a development official in Praslin, said: 'I have been diving for the last 40 years and I have never encountered any problems. Sharks are actually very rare because they are hunted for their meat and fins.' Seychelles Ambassador to the United States Ronald Jumeau told *ABC News* that reports of the killer shark's size being 2m (6.5ft) long were false: 'We haven't had a reliable sighting – we don't know where that number came from. People have been guessing. We have no idea whatsoever about what type of variety of shark it would be. He added that a special committee had been assembled, comprised of standing authorities from the National Parks, the Seychelles Fishing Authority, local police, the Coast Guard, marine biologists, hotel security and residents to coordinate their actions in guarding the area and preventing a third attack: 'A domestic advisory has been announced to make sure people don't go too far out. They are monitoring the waters to insure that they can get the shark. There's probably just one shark, but we are certainly taking precautions. There is no panic on the island, no wild shark hunt, although some of the popular beaches around that area have been closed and are being patrolled.' Meanwhile, the tourist board issued its own reassurance:

SHARK ATTACK IN THE SEYCHELLES

The end of July saw the sad and unfortunate death of a French visitor who lost his life in what was considered to be a shark attack. 16 August saw another man killed by sharks. The Seychelles has never seen shark attacks in the past; it was a given that the sharks in the waters of the Seychelles are friendly creatures. Hundreds of people dive every day: local fisherman dive for sea cucumber harvesting and visitors dive for fun. Shark encounters are a given daily occurrence as this is the Indian Ocean and it is infested with sharks. In fact, some people choose to dive for the sole purpose of swimming with the sharks.

Who would have thought that a shark attack would happen in the calm and beautiful Anse Lazio bay? Why Anse Lazio? What is changing in the environment that would cause such an attack on a human? Whilst a lot of questions need to be answered, the local authorities are gathering the advice of experts outside of the Seychelles, who can identify the source of the problem and rectify it.

There are differing opinions as to how this situation should be dealt with. Some argue for killing the shark. But what if there is more than one shark? Environmentalists, on the other hand, argue that we should protect the shark. Is there a good or bad way of dealing with this? One thing is certain, however. There are now two gaps in the world because of this incident. There should never be a third. All that can be done is being done, at the present moment. If anyone has any advice or ideas as to dealing with this situation, now is the time to speak.

As for the current situation on the beach, it is now open for visitors. This is good, because the beach is one of the main attractions that visitors come to see. Swimming, however, has been banned until the water is verified to be safe. There is a 24/7 police patrol on the beach and they do not have an easy task. There are still people who want to swim with a shark that has killed two people.

On 18 August, residents and tourists on Praslin attended a beach ceremony in remembrance of Ian Redmond. A priest in flowing robes scattered petals from Seychelles' tropical flowers onto the water. There was song, sadness and stunned disbelief from all those present. One attendee said: 'Someone was playing the guitar and people were singing. There were around 120 people there, which is an enormous turnout for this island – that's quite a large proportion of the population. There must have been people who travelled for some way to attend. Many were visibly upset, holding handkerchiefs and crying. It shows how bad everyone feels for what has happened. They are taking it very seriously here.'

During the ceremony, a helicopter flew overhead, still scouring the waters for the killer shark.

On 22 August, a team of shark hunters, including Jeremy Cliff – who would later tell the authorities to close the beaches – was flown in from South Africa to track down the Great White, which already had a £2,500 ($3,000) bounty on its head put up by those desperate to retain the Seychelles' annual 19,00 British visitors. 'It is the most beautiful beach – if you see it, you'll know that, for sure. But business will be bad if the shark is not caught. About 95 per cent of our clients are foreigners, we rely on them – we have to find the shark and put an end to the problem,' said James Lepair, assistant manager at a popular Creole restaurant. Another local added: 'The shark is the enemy.'

The bounty prompted a group of 'shark vigilantes' to take action, with locals taking up harpoons and trawling the waters in their boats. One, Albert Dienville, vowed: 'He will not get away from me – I have been fishing these waters for 40 years and I know how to kill sharks. If he gets past the nets and the hooks and comes for my bait, I'll get my revenge.' Deep-sea fisherman Jean-Baptiste Pool, 49, said: 'I know how sharks think – and I know he will come back here soon. Now he has the taste for blood, he believes he controls this territory but when he comes back, we will be waiting. We have to kill him because the tourists will never come back as long as he is out there: we're in a fight for survival.'

It was all-out war, with a grey Navy warship on hand, ready to fire at the shark if it came into sight. Overseeing the operation, deputy regional police commander Sergeant Mike Menthy said: 'This creature has taken two lives and could take many more livelihoods unless it is caught and killed. No one wants to go in the sea with that thing out there.'

Some, however, did not want the shark to be killed. Discussions were opened on internet forums, including the following:

> I'll take this moment to convey my deepest condolences to Gemma Houghton, wife and loved ones of Ian Richmond and also to the family and loved ones of Nicolas Virolle. I cannot imagine what they must be going through. Still, if the reports are true and the Seychelles authorities are hoping to catch this shark, then I have to object to this plan of action. I understand the need for the Seychelles Government to protect their tourism industry; the beaches are integral to tourism in Seychelles and declaring it safe to swim is something that is important. However, the reaction to kill the shark is mistaken as diving is also part of the tourism trade. Moreover, the extenuating circumstances (waste disposal by boats, overfishing, etc.) are being ignored. Instead of killing it, how about erecting a shark-spotting tower for the area and also remove the boats or at least ban waste dumping in the bay? It's a UNESCO site and everything should be protected, not just what is convenient.
>
> For those who agree with me, please write to the Seychelles Government and tourism board to tell them how you feel. I certainly will not consider the Seychelles again if they did kill this shark when there are so many other ways to deal with this problem.

Meanwhile, the shark situation was certainly concerning the professionals. Jeremy Cliff said: 'The fact that there have been two shark attacks so close together suggests there is a serious problem and we must therefore assume it could happen again. It is always impossible to say for sure what is happening when we're dealing with sharks, but we must make educated assumptions based on the evidence we have.'

There was still some dispute over the shark's true identity, with some insisting the predator was a Tiger shark or Bull shark – both more commonly found in the Indian Ocean. Cliff's arrival on Praslin was reported by the Seychelles Tourist Board:

MARINE SAFETY PRESS RELEASE UPDATE

The shark experts, Mr. Jeremy Cliff and Mr. Michael Anderson-Reade from the KwaZulu-Natal Sharks Board, arrived in the Seychelles yesterday morning on an Air Seychelles Flight from South Africa. In the afternoon they met with senior government representatives from the Seychelles Maritime Safety Administration, Environment Department, Praslin Development Fund and the local police to discuss their work programme over the next 6 days. According to their Terms of Reference they will assist the Government of Seychelles with the following:

- Comprehensive prevention and safety measures
- Determine and propose a comprehensive list of prevention and safety measures that the authorities need to apply, along with the geographical extent and the length of time these will require to lower the possibility of a shark attack. These should include advisories to swimmers and divers, surveillance systems, rescue respond measures and first aid accessibility.
- Elimination of the potential danger
- Identification of the shark species and size from the information available to determine what may have

triggered or caused the attacks. What methods can be utilized to abate or remove the particular shark threat?

- Other contributing factors
- Identify and list other biological and anthropogenic contributing factors that may need to be considered, such as disposal of food from yachts and other pleasure boats from the shore that may affect shark feeding behaviour. How can these be best addressed?

The two South African experts also did some preliminary assessment of the injuries of the two victims from photos provided by the police and the piece of tooth that has been retrieved from the second victim by doctors. Based on the preliminary assessment of the injuries, they are inclined to think that the species of shark involved is a Great White, although they are not discarding the possibility of a Tiger shark at this stage. Further detailed studies will be conducted during the coming days.

This morning they have been meeting the local fishermen and in the afternoon they are meeting with local technical experts in an effort to gather as much information as possible about Seychelles maritime policies, laws, local marine conditions and climate.

On Mahé senior government officials met with the representatives of tourism businesses today and briefed them on actions being undertaken.

The ban on swimming in Anse Lazio, Petit Anse Kerlan, Anse Georgette, Curieuse and St. Pierre is still on. In other areas, swimmers and divers are being asked to take necessary precautions.

> In an effort to increase surveillance, support has been sought from Air Seychelles pilots, who regularly fly over the area to keep a lookout for any large fish in the sea and to report any sightings. This is being supported by patrol boats from Seychelles National Parks, local fishermen and the Coast Guard.
>
> The Government of Seychelles wishes to reiterate that it is against putting a bounty on sharks and the indiscriminate killing of sharks. It wishes to reaffirm the fact that it will apply internationally accepted best practices to address the problem based on advice from the shark experts.
>
> In the meantime all fishermen who catch a shark are being asked to report to Mr. Rodney Quatre, the research manager of the National Parks Authority, who is collecting data on sharks.

The shark attacks took a sinister turn with the realisation that the beast from the deep was following an astral pattern. Local fishermen noted Nicholas Virolle was attacked and killed two days after a new moon and Ian Redmond two days after a full moon. Now they awaited a third attack. Daryl Green, the fisherman leading the hunt, said: 'The next high tide like that one is around August 29 – that's when we expect the shark to come back and we are doing all we can to catch it.' The fishermen placed 250 baited hooks around the area where Ian Redmond was killed. Said one: 'We have to catch it. I'm not giving up – I'll be out there every night and so will others. Everyone is afraid of the damage this can do to the Seychelles.'

Slowly, the beach at Praslin was losing its reputation as one of the world's most stunning locations and some of those monitoring the horrific incidents wanted to put their own experiences in the blue-jewel waters forward in a realistic way. Writing in the *Telegraph* on 17 August, Tim Ecott described the island and its merits:

> Anse Lazio is not just a spectacularly beautiful beach; it's one of the best beaches in the whole of the Seychelles for swimming. Many of the beaches around Mahé and the other popular holiday islands are beautiful to look at, but the sea is often too shallow to allow swimming except at high tide. But Anse Lazio is picture perfect: there are no hotels, no buildings visible from the sea, and at either end of the crescent of pure silver sand there are the great sculpted granite boulders that give the Seychelles their unique appeal.
>
> The sand is soft, the bottom of the bay is usually clearly visible through the calm blue water and there are rarely any big waves or surf to challenge swimmers. Thousands of holidaymakers are taken there on day trips from Mahé or from the dozens of small hotels dotted around the coast of Praslin. For 30 years tourists have regarded a visit to Anse Lazio as one of the highlights of their visit to the island – second only in popularity to a visit to the nearby Vallée de Mai, the World Heritage Site where the coco-de-mer nut is found.
>
> For honeymooners Praslin is a delight, as is swimming at Anse Lazio followed by refreshing cold drinks and seafood salads from the delightful Bonbon Plume restaurant tucked behind the tree line. I have made over

500 dives in the Seychelles and have dived around Mahé, Praslin and La Digue many, many times, as well as around the further-flung islands like Desroches, Denis, Frégate and even at the very remote Aldabra atoll.

I have seen sharks in the Seychelles but only at specific sites, where commonly seen reef species like black-tips, white-tips or – occasionally, the slightly larger grey-reef sharks are a bonus to the underwater naturalist. But divers often lament the relative scarcity of sharks around the islands.

Last month I snorkelled at Anse Lazio. I didn't dive there because it wouldn't occur to me to use scuba equipment in the bay where there is generally very little marine life to see.

I swam out from the beach where Ian Redmond died about a hundred yards – far enough to get away from the other swimmers and snorkellers so that I could enjoy the fish life without interruption. The water was balmy – around 28°C [82°F] and there were parrotfish and sergeant majors flitting between the boulders that line the edge of the bay. At one point I spotted a flash of colour in the deeper water and swam out to investigate, discovering a school of baby squid swimming rank and file together.

Two weeks ago news of the shark bite that caused the death of the French tourist Nicolas Virolle came as a complete surprise to me. None of my friends in the diving industry in the Seychelles had ever heard of a bite on a swimmer, the only locally known accident having involved a Seychellois fisherman who was bitten by a tiger shark while hunting for turtles in deep water in

> 1963. People working at Anse Lazio were shocked by the first shark attack and no one who worked at the beach could ever recall a shark having ever been sighted nearby. A second incident was unimaginable.

It was down to the experts to ascertain what had made a rogue shark carry out the two attacks, but they had little to go on. The attacks were the first fatal ones around the Seychelles for nearly 50 years. Australian research scientist Dr Jonathan Werry told Sky News on 17 August: 'Sharks have specific drivers, generally physical or biological drivers that coincide with their movement into a coastal area. And in many cases, when you look at an attack there are other features that have correlated and led to that attack.' One of these features may have been the lure of a readymade meal for sharks – thanks, once again, to the fishing industry. For the Seychelles is home to the biggest tuna canning factory in the world and the dumping of fish waste attracts many a predator, despite the ban on discarding rubbish close to the shore and the reinforcement of a ban on dumping waste from yachts and other vessels into coastal waters. These measures were put in place with the sole aim of keeping sharks away from their abnormal habitat of the shallow areas – and humans.

Some reports suggested this particular killer shark had been circling the waters for more than six months. Jeremy Cliff believed a change in the shark's environment may have caused the attacks, saying: 'Perhaps there are issues with the sea bed or reef. Changing conditions could attract sharks, too. Dealing with a problem like this is not as easy as just catching the shark: something must have changed their behaviour and brought them to shore, and we need to find out what.'

On 30 August the funeral of Ian Redmond was held at St Michael and All Angels Church in Dalton, Skelmersdale – the same family church where he and Gemma had married. Hundreds of people attended, all mute with disbelief at the tragedy. Held tightly by her parents, Gemma followed her husband's coffin and a lone bagpipe player led the sad procession. Once more she paid tribute to the man she had lost in such horrific circumstances in a truly heart-breaking speech:

A LETTER TO MY HUSBAND

To my darling husband Ian, I love you with all my heart, I always have done and I always will do.

It is so hard to express just how much you mean to me, but I so want to try. I want to respond to the beautiful words that you put into your card and speech on our wedding day.

When we first met each other almost nine years ago, I remembered being overwhelmed by a desire to become a part of your world. You were so full of life, exciting, fun and interesting, and you remained so until the day we lost you.

I have always felt so lucky that you chose me to share in your life experiences and I am so grateful for some exceptionally special memories and times.

I remember our first date at Manchester Art Gallery. We held hands constantly. I always loved the feeling of entwining my fingers between yours.

I remember so many happy times spent climbing and

bouldering. You would climb with your friends and I would watch. You would always check if I was OK, and I always was because I was watching one of your many talents with pride – I was in awe of how fearless and enthusiastic you were with so many things.

I remember the extensive interests we shared, like art and films and the outdoors. Days out to places like Chatsworth sculpture exhibition, the cinema and the Lake District, which may have seemed ordinary to others, I treasured because it meant that I had found someone like me. You were my soul mate and my best friend.

I remember seeing you kneeling on my doorstep the night that you proposed to me. You had a single red rose, a bottle of champagne and a giant engagement ring that you had constructed from wire and tinsel. You melted my heart and your actions epitomised your thoughtful and caring nature.

I remember the exciting times that we spent working on our new home together – painting, lifting flagstones, demolishing porches, demolishing rooms – any skill that you put your mind to, you achieved.

You were so intelligent, with a great eye for detail. Your dedication to make our home special and right and comfortable for us knew no bounds.

I want to thank you for working tirelessly on our perfect home. I promise I will finish it for us and live there with you in my heart.

I remember how you learned to dance with me for our wedding day. You were so good at it and put 100% into it – just like you did with everything. When we danced, you never took your eyes off me for a second. Thank you – I felt so much love for you.

I remember our beautiful wedding day. It was quite simply the most wonderful day of my life. You were so much more than I could have ever wished for in a husband and memories from that day are just some of the richest and warmest I will have of you.

I remember the precious times we shared on our honeymoon. We were having such a happy time. It was a lovely adventure and we were enjoying experiencing new things.

We felt so at peace, so relaxed and were so excited about the future. It was truly romantic. Each and every time you went swimming or snorkelling and I watched you, I could never quite believe that I was married to you.

You were the most handsome and perfect man I have ever seen and I always longed for you to come back out of the sea and be back with me.

Ian, there just aren't enough superlatives to describe you. At work where you excelled and dedicated so much of your time, they described you as the 'gold standard' of developers. But you were also the 'gold standard' of men – always smiling, always finding time for the family and friends that you loved so dearly. You were never cross, just kind.

Above all you were the 'gold standard' of husbands: amazing, courageous, inspiring, entertaining, patient and loving. You gave me such confidence and strength. You filled me with such pride.

I've said it before and I'll say it again: you were (and

are) my best friend. We could talk about anything and everything with each other – laugh, cry or just be quiet. You completed me and are the best thing that has ever happened to me.

Whilst I cannot believe that you are gone – I am in shock and hurting so very much – I am comforted and consoled by the rich tapestry of memories that we formed over our nine years together. Thank you.

I promise you will never ever be forgotten and I will miss you so very much. You, above all people, knew how hard I found it to say goodbye to others and so I know you will understand if I don't say it. What I will say instead is "See you soon" – I'm looking forward to it so very much.

All my love now, forever and always,

Your wife, Gemma.

Incredibly, Gemma was still able to extol the virtues of the islands where she had endured such a nightmare, saying: 'The last thing I would want is for any of these events to affect the Seychelle Island people, their livelihoods and the tourism in the area. It's a beautiful area, people must come. It's a one-off accident and I know that everybody is doing everything they can to ensure that the islands are safe – the restaurants on the beaches and the places on the beaches and the hotels shouldn't be affected by it.'

On 1 September, local officials agreed to Jeremy Cliff's recommendation that anti-shark nets should be put in place off Anze Lazio beach. Until this was done, the ban on swimming remained in place and the hunt for the rogue shark was still on.

SHARK NETS INSTALLED ON ANSE LAZIO BEACH ON PRASLIN

Information was received from Mahé over the weekend that the anti-shark campaign continues to be high on the country's agenda following two unprecedented shark attacks some weeks ago.

While fishermen and coastguard are still searching for sharks in the pristine waters off the Seychelles inner islands, in particular around Praslin, where the attacks took place, and have in fact landed several of the beasts since the hunt opened, more protective measures have been instituted by the government, hand in hand with the affected tourism industry. Boat patrols continue to guard the beaches off Praslin to ensure that predator fish are spotted early and destroyed.

It was learned that an anti-shark net has been installed off the famous Anse Lazio beach on Praslin and is now only awaiting inspection by the Seychelles Maritime Safety Administration before swimming can be officially allowed again in certain sections of the beach. Other beaches too will then see the installation of similar nets, aimed at keeping the predators away from the shallower waters near the beach where tourists like to swim and snorkel. Praslin, alongside Mahé, the archipelagos largest island and home to the capital Victoria and La Digue, are the most popular tourist islands and of crucial importance to the country's tourism industry.

Although accepted as a necessary safety measure, the nets were regarded as a blight on the stunning beach outlook. South African shark specialist Jean Pierre Botha said that although the nets were highly effective, 'the drawback is that they have a tremendous negative impact on the ecology, catching and killing many protected species including whales, turtles and protected species of shark.'

It was now time to admit that mistakes had been made in the hunt for the shark which had brought terror to the Seychelles. Jeremy Cliff changed his original opinion on the shark species. It was probably not the greatest predator of the sea, the Great White, but more likely to be a large Tiger shark – a common inhabitant of the Indian Ocean, and not, as local authorities claimed, 'a foreigner' to their waters. Said Cliff: 'From close examination of photographs of the injuries, it appeared that large Tiger sharks in the region of 4m [13in] were responsible for both attacks. It is impossible to confirm that the same shark was responsible, but it cannot be excluded.'

In a bid to end the terror, local fishermen caught around 40 sharks. This sad haul included Grey Reef, Snaggletooth, Lemon, Nurse sharks and Blacktip Reef sharks – all described as 'posing no major threat to humans'. Among them was a Tiger shark, whose stomach contents were examined for human remains – and a wedding ring. Anyone catching a shark was ordered to report to the National Parks Authority, collectors of shark data. (Shark meat is often eaten in the Seychelles and the fishermen involved would quite possibly have been fishing for sharks anyway on Praslin or Mahé.)

In the end, the experts had to confess they could find no reason for the attacks; the only common link being that they all took place late in the afternoon. Said Cliff: 'In general, attacks in the late afternoon are not unusual as this is when sharks, which are generally most active at night, tend to start moving inshore to hunt for food.'

The total ban on swimming off Anze Lazio beach was lifted, but swimming was still only allowed within the perimeter of the exclusion net. For now, all was returning to normal at the holiday paradise. Meanwhile, the saga of just which shark had killed the two tourists continued, though. Both earlier identifications had been wrong, it seemed. In October 2011, yet another official statement was issued:

> According to a communiqué from the Ministry of Home Affairs, Environment, Transport and Energy (MHAETE), the shark tooth fragment found in Mr Redmond's leg and sent to a genetics expert in Florida has been identified by DNA analysis as belonging to the species *Carcharhinus leucas* (bull shark).

This final – and hopefully conclusive identification – was made by Professor Mahood Shivji, director of the Save Our Seas Shark Center and Guy Harvey Research Institute at Nova Southeastern University in Florida, with the help of a colleague: Dr. Larry Grillo, a dental surgeon in Miami. The two men retrieved a small amount of dentin from the tooth fragment, which they said yielded sufficient DNA for analysis. Another member of the team, Chris Clarke, said a Bull shark was one of the original suspects, together with a Tiger shark and a Great White: 'It's really helpful to know which shark species was involved. We can now focus our attention on studying the biology and movements of the species in Seychelles waters to help local managers prevent future incidences, as well as ensure the safety of sharks as critical components of a healthy marine ecosystem.'

The MHAETE announced: 'For the Government of Seychelles this is a major breakthrough because it pinpoints to the identity of the shark species involved in the attacks. With this knowledge, the government can now design and implement a targeted research programme to better understand the population dynamics, behaviour and distribution of Bull and Tiger sharks within our coastal waters.'

Local marine scientists and environmental experts from the Seychelles met to finalise details of a research project to be implemented over the next four years with Chris Clarke and the KwaZulu-Natal Shark Board.

Seychelles interior minister Joel Morgan expressed his 'satisfaction and gratitude' to Dr. Clarke and Professor Shivji and his team 'for providing such groundbreaking and world-class services free of charge to the Government of Seychelles on such a critical issue.'

For now, it was safe to go back into the water…

THE BULL SHARK

Also known as Zambezi shark, Van Rooyen's shark, the Ganges shark, the Nicaragua shark, Swan River Whaler, shovelnose, slipway grey and square-nose. The Bull shark is one of the three species most likely to attack humans. It has a tendency to headbutt its prey. Sharks must keep salt in their bodies to survive and most can live only in saltwater but Bull sharks have developed special adaptations – the way their kidneys function and special glands near their tails – that help them retain salt in their bodies even in freshwater. Scientists are still studying these sharks to figure out why they developed this unusual ability.

Description: Bull sharks take their name from their short, blunt snout. They have thick, stout bodies and long pectoral fins. Grey on top, they are white below with grey-tipped fins (particularly seen on the young of the species).
Size: Adult males can be up to 7ft (2.1m) long and weigh from 90–230kg (200–500lbs). Females are larger.
Diet: Just about anything, including fish, dolphins and other sharks.
Habitat: Warm, shallow waters of all the world's oceans and tropical shorelines, but also often the brackish waters of estuaries and bays (where they may attack humans out of curiosity).
Seen: Spotted along the Mississippi River, the Amazon River in Peru, Lake Nicaragua and Africa's Zambezi River, as well as more regular haunts of the western Atlantic to southern Brazil, eastern Atlantic (Morocco, Senegal to Angola), Indo-West Pacific (Kenya and South Africa to India), Vietnam to Australia, the eastern Pacific (southern Baja California, Mexico to Ecuador).

THE TIGER SHARK

The second-largest shark in the world (beaten only by the Great White), these large, blunt-nosed predators have truly earned their reputation for being man-eaters. They are second only to Great Whites in attacking humans but because they have an almost completely undiscerning palate, they are unlikely to swim away after biting a human, as Great Whites frequently do. Solitary creatures, they have amazing eyesight that enables them to hunt at night. They can also change their colours from a blue to a green to help blend in with their surroundings. The Tiger shark is heavily harvested for its fins, skin and flesh and the livers contain high levels of vitamin A, which is used in making vitamin oil. They are included on the Threatened Species List.

Description: Tiger sharks are so named because of the dark, vertical stripes found mainly on the young of the species. As these sharks mature, the lines begin to fade and almost disappear. The shark has a blunt nose and sharp, highly serrated teeth and powerful jaws that allow them to crack open the shells of sea turtles and clams.
Size: 3.25–4.25m (11–14ft) and weighing up to 635kg (1,400lb). Larger specimens have also been found.
Diet: Consummate scavengers. The stomach contents of captured Tiger sharks have included stingrays, sea snakes, birds, squid, car licence plates, old tyres, jewellery, clothing and bits of ships and boats! No wonder it has gained the nickname 'wastebasket of the sea'.
Habitat: Coastal waters close to shore of the outer continental shelf and offshore, including oceanic island groups.
Seen: Off the Atlantic coast of the US, Tiger sharks may be found from Cape Cod, Massachusetts, to the Gulf of Mexico and Caribbean Sea. Off the Pacific coast, they appear from southern California southward. In the western central Pacific, Tiger sharks can be spotted in the Hawaiian, Solomon and Marshall Islands.

GREAT WHITE SHARK

Also known as White Death, White Pointer and White shark. Great White sharks are at the very top of the marine food chain: the most aggressive inhabitants of any sea and fearsome predators. They can smell a single drop of blood from over a mile away. Prey can run, but they cannot hide for the Great White can detect and home in on small electrical discharges from hearts and gills. Of all the species, it is the only one that will bring its head above water, probably to search for large prey. To date, this shark has been responsible for 63 deaths worldwide since 1876, with 232 recorded other attacks.

Description: A robust, torpedo-shaped shark. The upper and lower lobes of the caudal fin are about even in size. Despite its name, it only has white on its underside and can be grey, black or blue-coloured on top. It has serrated, triangular teeth which are almost symmetrical. An adult has at least 240 teeth at any time and can grow them when needed, but they also lose many teeth every day.
Size: On average, around 5m (16ft) long, but can grow up to 6.6m (22ft). The female grows larger than the male. These sharks can weigh up to 525kg (1,157lb). The largest Great White ever caught was officially recorded as being 6m (19.5ft) long.
Diet: A solitary predator who preys on a variety of fishes and marine mammals (most commonly salmon, hake, halibut, mackerel, tuna, harbour porpoises and harbour seals), but it will also eat other sharks, sea turtle, seabirds, seals and the blubber of dead whale carcasses. It can smell a seal colony from 3.2km (2 miles) away.
Habitat: Coastal and offshore waters as shallow as 1m (3.2ft) of the continental shelf, but will sometimes stray into bays and harbours. Frequents both surface waters and waters as deep as 1,280m (4,199ft).
Seen: In temperate, subtropical and tropical waters world-

wide. Western Atlantic (Newfoundland to Argentina, including the Bahamas), Eastern Atlantic (France to the Cape of Good Hope and the Mediterranean Sea), Eastern Pacific (Gulf of Alaska to Chile), Central Pacific (Easter Island, Hawaiian Islands and Marshall Islands), Western Pacific (Siberia to Tasmania), Red Sea and Indian Ocean, including South Africa and Mozambique, Madagascar, Mauritius and Seychelles, and Western Australia.

SHARK INHABITANTS OF THE INDIAN OCEAN

There are at least 17 kinds of shark living in the Indian Ocean:

- Blacktip shark
- Bull shark
- Oceanic Whitetip shark
- Dusky shark
- Great White shark
- Tiger shark
- Bluntnose Sixgill shark
- Shortfin Mako
- Megamouth
- Blue shark
- Whale shark
- Scalloped Hammerhead
- Great Hammerhead
- Spiny Dogfish
- Kitefin shark
- Thresher shark
- Pygmy shark
- Smooth Hammerhead.

CHAPTER 3

ROGUE SHARKS OF REUNION ISLAND

'WE SAW A CLOUD OF RED BLOOD IN THE WATER WHERE HE HAD BEEN...'

It was described as a frenzied attack and lasted less than 30 seconds, but in that time champion surfer Mathieu Schiller died in the jaws of a shark. He was one of several victims of the rogue sharks of Reunion Island.

Schiller, 32, was dragged off his surfboard in the waters off Boucan Canot beach on the French-owned island of Reunion in the Indian Ocean. His death was a random choice by the shark – for Schiller was just one of a large group surfing in the sea and its target could have been any of them. Said one witness: 'There were around 20 people in shallow water and around five surfers out deeper when it happened. We saw the shark's nose emerge and the man just vanished. It was very sudden then the animal just swam off. Some of those nearby tried to reach him, but his body was dragged away by the current.' A fellow surfer said: 'I was waiting for a wave with my feet in the water. We saw the shark and the next thing I knew was that he screamed and raised his arm. We then saw a cloud of red blood in the water where he had been.' Another witness noted that the shark had attacked with 'terrifying speed.'

Reports of what happened after the attack on 19 September 2011 were conflicting and confused. One said that the surfers managed to get Schiller back on his board despite the gaping wound in his leg. Others noted the shark had become frenzied after blood hit the water, returned and hit Schiller's surfboard, knocking him into the water. Yet another stated that a wave hit the board and Schiller fell off and disappeared beneath the surface. Further, rescuers searching for Schiller's body reported a 4m (13ft) Bull shark had struck their dinghy, trying to overturn it; others said a Tiger shark was the attacker.

Despite a search by boats, jet skis and a helicopter while his family and friends watched on from the beach, Schiller's body was never found. It was believed to have been carried away in the waves. Before his death, lifeguards had hoisted a red flag to indicate bathing was banned because of the rough sea conditions but it was claimed the surfers had ignored the warning. The rough sea also hampered rescue attempts in the dinghies. But Schiller and his friends were defended over their decision to go into the water, with one member of his family saying: 'Red flags are for swimmers, not for surfers. The world champion of Body Board in an interview with the French TV news in the Island of the Reunion stated that the water was clear this day, and not different from any other good surfing days so to say that Mathieu Schiller should not have been out is ridiculous. He was a professional of this sport and in no way a daredevil. Accidents are unfortunate and the local administration needs to implement protection by installing nets like they have done in the Seychelles after another fatal accident. Please don't tarnish this young man's memories by putting the blame on him. It only hurts his family more, is irresponsible and does not serve anything or anybody.'

Following the attack, all water activities were banned outside a limited area and mayor Michel Lalande announced: 'This repetition of deadly attacks requires public authorities, in consultation with the mayor of St Paul, to take emergency measures. There must be no question of exposing both professionals and our population to any risk.' An investigation was ordered into the species of shark circling the waters, with Lalande simply stating: 'This is not a shark hunt – I am relying on professional experts.' He expressed his 'sincere condolences' to Schiller's family, adding: 'We cannot forget that he was an instructor in a surf school and knew the sea and its dangers.'

A professional surfer (he was a former bodyboarding champion of France) and lifeguard, as well as a surf-school owner, Mathieu Schiller was the second person to be killed by a shark that year and the sixth to be attacked. In fact, Reunion Island, located in an isolated spot about 40km (24 miles) from the coast of Madagascar, has one of the world's highest numbers of shark populations and attacks on swimmers, surfers and divers. Since 1980 there have been 24 shark attacks off the shores of this island, with 13 of them being fatal. As one tour operator warned: 'While not many people head to this tiny island, those who do should be extra careful before they get in the waters here.'

Three months earlier, there had been another fatality. On 15 June 2011, local resident and bodyboarder Eddy Auber, 31, died after a shark attack at Boucan Ti, close to the beach. Witnesses reported seeing him raise his hand and then go twice underwater. Rescuers went to his aid but said they were thwarted by the sight of 'several fins', leading to the belief several sharks had been involved. When Auber's body was eventually brought back to shore, it was found to have two deep bites on his thigh and torso; his right arm had also been severed. A shark bite was clearly visible on his bodyboard. Said one observer: 'The body was very messed up – there were several bites. He was already dead when he was out of the water.'

It was the first fatal attack at this site, a popular one for surfers and bodyboarders because of the swell. One local said that at the time of Auber's death the cloudy waters, swelled by heavy rainwater, were 'particularly conducive to the presence of sharks.' Fanch Landron, president of the Squal'idees association which studies the relationship between sharks and humans, said that shark group attacks are not unusual and can be provoked by 'confusion over food.' The last reported sighting before this attack had been just ten days earlier when a Bulldog shark was seen off Saint-Gilles. It was also reported that in 2007 a young bodyboarder had been bitten at the same spot.

In September 2011 the head of the French overseas department of Reunion Island announced it had authorised the killing of ten sharks belonging to the 'most dangerous species' – Bull and Tiger sharks – following a series of attacks. This decision was part of a programme to reduce the 'shark risk' and had been taken after discussion with the local mayor. The sharks were to be killed by professional fishermen over three days, during which all water activities would be banned. A spokesman said: 'These sharks have settled in the area and we aim to create a disturbance in the shark population.' The high number of sharks was described as 'exceptional and hardly explainable,' but the killings provoked criticism from animal welfare groups, including the Brigitte Bardot Foundation, who described the measures as 'disproportionate', adding: 'How can we invade all areas and clear anything representing a threat to man?' These two species are not protected by French law but assessed as 'near threatened' by the International Union for Conservation of Nature (IUCN). Under the programme, any caught sharks under 1.50m long (4.9ft) and all others apart from Bull and Tiger sharks were to be tagged and released.

On 13 October 2011, a woman's body was discovered floating in the water close to shore at La Possession on the northwest side of Reunion Island. The authorities retrieved a leg and the torso, both bearing shark wounds. An autopsy confirmed the woman had been the victim of a shark attack around 25 hours before her remains were found. La Possession mayor Roland Robert banned all water activities until further notice.

Strangely, the Reunion Island tourist board state: 'There is nothing particularly exciting about Reunion Island's animal life, although there are few dangerous species. One should be wary of stonefish, centipedes and sharks.'

Just over a month later, on 11 November 2011, there was another shark attack off Reunion Island. Diver Jean-Paul Delaunay, 42, was with fellow divers less than a mile from the shore in the waters of Anse des Cascades at Saint-Rose. He suffered several injuries. One of those accompanying him, Ghislain Mussard, said: 'We were all four under the water. Jean-Paul was about 15m (49ft) from me. When we got back to the surface, he said that he saw a shark and was attacked.

At the edge of the Zodiac boat, we found that he had been bitten on the left foot. He has lost his palm, probably torn or shredded by the shark. There was a big gash and he was bleeding a lot. Jean-Paul was the only one who saw the shark so we don't know if it was small or big.' A doctor among the dive party managed to stop the bleeding and Delaunay was taken to hospital in Saint-Benoît.

The year of 2011 saw several non-fatal but nonetheless nasty attacks off Reunion Island:

On 5 October 2011 Jean-Pierre Castellani, 51, fell into the water after a shark – believed to be a Tiger shark – bumped his canoe, close to Cap La Houssaye in Saint-Paul. He had a lucky escape and was not attacked. Castellani was in the water for ten minutes before a passing boat came to his aid: 'I protected myself by automatic reflex and hit the shark. When I found myself in the water, I did not know if the shark would come back again,' he said. His canoe was bitten instead.

'He had the fright of his life. Fortunately for him, the shark did not return. Not having been wounded, he did not need special assistance and was brought to shore,' said Nicolas Le Bianic, director of the CROSS (Regional Operational Centre for Surveillance and Rescue). A ban on all water activities at the beach of Boucan, near the Rivière des Galets, was put in place. A video was placed on the internet showing an encounter between a large Hammerhead shark and a fisherman in a small boat off Saint Gilles, Reunion Island.

On 6 July 2011,15-year-old Dussel Arnaud was attacked by a shark while surfing off the beach at

Roches Noire, off Saint Gilles. Incredibly, he escaped with only minor injuries to his face and ankle but his surfboard was bitten in two.

On 19 February 2011, despite losing his leg, 32-year-old surfer Eric Dargent survived the first-recorded shark attack off Saint-Gilles. He had only arrived on the island with his wife that morning and was enjoying the waters shortly before sunset. Several surfers were in the water when they heard a scream. One said: 'I thought Eric was stuck between the coral beds so I went to help him but when I pulled up alongside him, I saw the wound and the blood. I used a Lycra bodysuit to make a tourniquet and we brought him back as quickly as possible. He said: "It's a shark! It bit me and I have lost a leg." Then he more or less lost any knowledge of getting back to shore.'

A witness said the attack was 'unlucky' as the water had been clear and the waves 'beautiful', but added: 'Sharks are unpredictable and can attack anywhere.' Dargent had his left leg torn off at the knee but was conscious when he was taken to hospital, where he stayed for ten days. He later recalled the attack: 'I had not seen a fin or even any movement around my board. Suddenly, a shark attacked me from below and grabbed my leg. I struggled; I hit it. It then tried to pull me to the bottom. I resisted; it let go. Every day I can now do a little more, like go shopping and get my children to school, but I have pain. Certain movements are impossible to me and I have fallen many times. I had to relearn to walk. At first I could not stand my prosthesis more than ten minutes and I walked with crutches.' He would later return to the waters off Reunion, describing it as 'a real treat.'

According to the Observatory Marine Reunion, 30 shark attacks have been recorded around the island since 1972, of which 17 were fatal. Before the serial attacks of 2011, the last attacks were on 27 March 2010 at St. Benedict, when 34-year-old surfer Oliver Schorebreak escaped serious injury by battering a shark with his board. On 3 May of that year, Michel Touzet (59) also had a lucky escape. He recalled: 'Diving conditions were good. The time was 9.30am; water temperature 29°C [84°F]; rather good visibility with no current, few fish and no fishing boats. The sea bottom was approx 70m (229ft), a soft coral drop off. About 10m (32ft) from me, my buddy was taking pictures. We'd been underwater for 10 minutes at a depth of 45m (147ft) when the attack occurred.

'The shark probably came from the bottom. I saw it for only a fraction of second. Its big mouth was wide open before it hit me. I heard a big noise, waoomm, on the impact. I felt dizzy for a few seconds after it hit my head and chest. The attack was so sudden that I didn't have time to do something, not even to be afraid. I had to use my octopus to breathe since my regulator was torn off, the mouthpiece broken and my mask skirt punctured. I went to my buddy, who assisted me for the ascent following the usual procedure. During the ascent I made a compression on my left forearm to lessen the bleeding. None of the other divers saw the shark except for my buddy Philippe – they were approximately 30m (98ft) away. It was a "hit and run" since the shark did not come back, thank God.

'The wound on my arm was oddly almost painless. A deep cut in my left forearm muscle and one on my chest were sutured under local anaesthetic with Lidocaine [a numbing cream] in a small Malagasy clinic in Hell-Ville. Luckily I had no sinew or nerve damages, and no broken ribs. Since the accident I have dived again in Reunion without problem and haven't met any other sharks yet. Before all this I had very few experiences with sharks. I had met a few Whitetips, Grey sharks or Whale sharks. I wonder what advice I could give to anyone after this incident! I still love these splendid animals and look forward to meeting them again in a more peaceful way. After talking to fellow divers, it may have been a Tiger shark who tasted me, but no one can be sure.'

- On 6 October 2004 Vincent Motais (15, and France's Junior Bodyboard Champion) was attacked by what witnesses thought was either a Bull or Tiger shark as he was bodyboarding at the Conservatoria District, Petit Paris, near the town of Saint Pierre. His left leg was horribly savaged and had to be amputated.
- On 8 September 2000, Karim Maan, 27, was badly bitten on his left arm by a 3m (9ft 10in) Tiger shark as he was surfing at dusk at Pic du Diable, near Saint Pierre. Maan had been about to paddle when he and his board were suddenly lifted out of the water. He was then subjected to a frenzied attack. 'The shark continued to attack and to hit me. I saw its mouth very close, I could have kissed it,' he said. 'When I came up for breath, I saw its head coming at me and I heard a crunch as it bit my board and my arm at the same time.'

The shark chomped on Maan's surfboard, let go, but went on the attack again. Maan then hit it on the nose with his board – a tactic which shocked it enough for it to go away. 'I was saying to myself, "It will attack again – I'm finished, I'm finished,"' he said. But he was somehow able to get back on his surfboard, catch a wave and return safely to shore. Following this, he needed 20 stitches and although he said he would soon be back in the water, he would give Pic du Diable a miss for a while.

Sickening Shark Secret

Reunion Island has a dark and sickening secret involving its shark population. In 2003, reports began to circulate of live and dead dogs and cats being used as shark bait by amateur fishermen. This claim was backed up by the Paris-based Fondation 30 Millions d'Amis (30 Million Friends Foundation) and other animal welfare organisations. Reha Hutin, president of the Foundation, sent a film crew to the island to obtain evidence of this. The team took videotape and photographs of dogs with multiple hooks sunk into their paws and noses. Hutin said: 'It didn't take long for the crew to fine three separate cases. From then on, everyone started to take the whole story seriously and realised it was true.'

A vet successfully treated one of the dogs – a six-month-old Labrador pup with a large fishhook through its nose – at an SPA (Société Protectrice des Animaux, or Animal Protective Society) clinic in Réunion's capital, St.-Denis. Unlike most of the hooked animals, the dog turned out to be someone's pet, according to Saliha Hadj-Djilani, a reporter for the Foundation's television programme shown in 2005. The dog had apparently escaped its captors and was taken to the SPA by a concerned citizen. Fully recovered, the animal was returned home to its owners. The other two cases uncovered by Fondation 30 Millions d'Amis were strays, who were sent to new owners in France.

Although the Fondation planned to finance a sterilisation program on the island to reduce the stray overpopulation, Hutin said many locals viewed the strays as vermin: 'There's no value to the life of a dog there,' he commented. In September 2008 the first court case was held, involving an amateur fisherman charged with using live dogs as bait. Authorities had found a seven-month-old puppy at the home of John Claude Clain; the animal had three fishing hooks in his paws and nose. Clain, a 51-year-old bread deliveryman, denied intending to use the dog for bait, but was found guilty of animal cruelty and fined £5,000 ($8,000).

Animal welfare groups said the case was not an isolated one. Fabienne Jouve of GRAAL (Groupement de Réflexion et d'Action pour l'Animal) based in France said: 'Lately, almost every week on the island one dog has been found with hooks, not counting the cats found on the beaches partially eaten by the sharks.' Once fishermen capture the animals, she said, the dogs and cats are hooked the day before being used as bait, 'so they can bleed sufficiently.' Some escape before being tossed into the ocean; others are not so lucky. After hooks are plunged into their paws and/or noses, the animals are attached to inflatable tubes with fishing line and reportedly dumped into the ocean – normally at night to avoid detection. In the morning the men return to see if a shark has been caught. 'Barbaric practices have no excuses whatsoever in the twenty-first century,' said Jouve. At one point, the Sea Shepherd Conservation Society was offering a $1,000 reward to any Réunion police officer who arrested anyone using live dogs or cats as bait for sharks.

The Embassy of France in Washington maintained although the practice was not 'unknown,' its 'occurrence and acceptance is not nearly as prevalent as news reports would make it seem.' In response to an avalanche of protests about the cruelty, the involvement of several animal welfare groups and online petitions, the Embassy issued a statement:

We too denounce the barbaric practices. Such acts are obviously illegal and will not be tolerated on French territory. But while we share your revulsion, we would like to emphasize that the practice of using live dogs or cats as shark bait is, in fact, exceptional and isolated. It was never widespread nor traditional but introduced by some ruthless individuals and has been strictly banned for decades now.

TV reports that raised initial indignation when they were aired in France and abroad in 2005 were filmed locally in 2003 following the discovery of a mutilated dog. The last few months have seen two identical events which received heavy media coverage (one of these events was soon determined to be a false alarm). But can these vile occurrences lead us to conclude that there is an ongoing tradition of barbarism on Reunion Island?

Reunion Island, a French territory and a European region, obeys the laws and regulations of the French Republic and the European Union. It respects the rule of the law and does not practise inhumane ancestral practices. The facts that elicited a complaint are the act of a few isolated, irresponsible parties who are being sought by the police and will be brought to justice. All suspicions of such acts will be investigated, and animal protection organizations that have any specific information on these matters are strongly encouraged to inform French police authorities.

The French minister for agriculture and fisheries, Dominique Bussereau, is fully aware of the media and public outcry regarding this issue and has written to the French National Assembly to emphasize that several measures have been taken to strengthen already existing laws. Veterinarians have been directed to immediately report any suspicious wounds to authorities and the police will increase their inspection of fishing and pleasure vessels. Meanwhile, a sterilization campaign, launched in 2001 to reduce the number of stray dogs and cats on the island, continues.

Animal rights are an important issue in France: over

half of French households have at least one pet, and France has some of the world's most stringent animal rights legislation. French law provides for the prosecution of those who are cruel to animals. Voluntary cruelty to animals is punishable by a sentence of two years in prison and a 30,000 Euro fine (equivalent to about $36,000).

CHAPTER 4

SHARK DEVASTATION DOWN UNDER

'I HELD WHAT WAS LEFT OF HIS LEG TOGETHER...'

Texan George Thomas Wainwright, 32, died of 'shocking' injuries in a Great White shark attack as he dived at Little Armstrong Bay, Rottnest Island, off the coast of Western Australia, on 23 October 2011. The creature was reported as being about over 3m (10ft) long.

Wainwright's disappearance was not noticed until two people on board the dive boat saw bubbles coming from the water – indicating air had escaped from his breathing equipment. Moments later, his body rose to the surface. Water police officer Greg Trew said the diver had surfaced in a 'flurry of bubbles' and was horrifically injured. As he was hauled onto the boat, he was believed to be dead. 'It was a cloudy day, which is the same as we had the other day at Cottesloe, and they are the conditions sharks love,' said a Western Australia police sergeant, referring to 64-year-old swimmer, Bryn Martin, who was fatally attacked on 10 October as he glided through the waters off Perth's Cottesloe Beach. All that could be found was a pair of torn swimming shorts bearing teeth marks of a Great White. A police spokesman commented: 'Our information on a preliminary examination was that the tearing was consistent with that of about a 3m [10ft] Great White shark. What that information tells us is that our concentrated efforts on this search now are recovery as opposed to the hopes of finding Mr Martin alive, which sadly, have faded.' Indeed, Martin's body was never recovered.

The fatal attacks were among a series over recent years off the beaches of Australia.

- On 5 September 2011, 21-year-old Kyle James Burden was killed south of Perth by a Great White – his lower torso had been torn away. Friends who witnessed the attack said the shark lunged up from underneath Burden as he swam on a bodyboard. A police spokesman said there had been no warning of the attack: 'Out of nowhere it would appear the young fellow has been taken by a shark; nobody actually saw the shark itself but they retrieved the young man's body in the water amongst some blood. You could have described it as perfect shark conditions – you know, dark and gloomy water, overcast skies, light rain falling... There was whale action earlier in the morning in the bay and there were seals about.'

The incident sparked yet another debate over whether the killer should be hunted down and destroyed. One of Burden's friends commented: 'Kyle was a bit of a greenie – he did like nature and he got in touch with nature. But I can recall him actually saying that those big sharks are a threat to our community. He would back killing them.' A local official agreed, saying: 'A lot of people say the water is the shark's territory but I think if they can find the shark responsible, they should get rid of it. If they have attacked a human in one of those areas they may want to do it again and I think we should be stopping that.'

Fellow surfers felt the same. According to one: 'We've been told that the shark that killed the man is an old female White Pointer [Great White] looking for easy prey. It's got to be caught, that's all there is to it.' Surf school owner Keith Halan said the Department of Fisheries had a duty to do something: 'These sharks are on the top of the food chain, a protected species, and are breeding in large numbers. I think it's ridiculous that we're allowing people to be taken here.' But in a poll in *Perth Now*, readers were against a shark hunt, with 52 per cent of 1,000 readers voting against any culling. A further 25 per cent believed shark nets were a better deterrent to shark attacks in Western Australia – a deterrent in force on the country's east coast beaches. The Department of Fisheries said it had no intention of killing the predator. Scientists also warned against an 'overreaction' to the fatal attacks. Barbara Weuringer from the University of Western Australia observed: 'It sounds a little bit like taking revenge, and we are talking about an endangered species.' Fisheries Minister Norman Moore added: 'This is a unique set of circumstances and I am desperately praying this is not the beginning of a new trend and we are going to have these on a regular basis.'

John West, curator of the Australian Shark Attack File, was in no doubt that Great Whites were responsible for all three attacks. He said: 'This species is known to inhabit the shallow waters along this coast and are known to migrate south around this time of the year to the seal colonies on the southwest coast. While they may stay around seal colonies – their natural prey – for months, they are not noted for sitting off a beach waiting for food to turn up. They are mostly individual, transient, inquisitive animals that will investigate objects in the water. Swimming, surfing or diving alone near aquatic animals, including seals and dolphins, far from the beach early in the morning or late in the evening may well attract a curious shark and increase the risk of encountering one. As the population increases and water-related activities become more popular, the number of people who go into the water every day also increases but the chance of encountering a shark still remains very low.'

Despite fears of a rogue killer shark, experts believed different species were most likely to be responsible for the attacks. Dr. George Burgess of the University of Florida (whose expertise was called upon following the Red Sea shark attacks late in 2010) said he believed the fatal incidents came as sharks were chasing whales: 'The chance of an individual shark being involved in all the incidences is extremely low as they travel 40 to 50 miles a day.' He added that there could be several reasons for the abnormally high number of attacks: the shifts in migration patterns due to climate change could be bringing sharks and humans closer together, while warmer sea temperatures were enticing more people into the water. 'This is a very unlikely candidate, one which has stayed around and got the taste for humans, particularly in the case of the White shark, which is so highly migratory,' said Dr. Burgess.

Western Australia Premier Colin Barnett said the sharks responsible would be hunted down and killed, if possible. To this end, fisheries officers set tuna-baited hooks off the island. This was an extraordinary move for it was the first time authorities had been granted permission to break the law protecting Great Whites, an endangered species. The exemption came about because it was now felt that it was more crucial to protect the public. Barnett added that the government would consider culls, responding to locals' complaints that shark numbers were increasing off the busy beaches in one of Australia's fastest-growing population areas. Referring to the three latest recent attacks, Western Australia state Fisheries Minister Norman Moore said: 'This is a unique set of circumstances, and I'm desperately praying this is not the beginning of a new trend and we're going to have these on a regular basis.'

New anti-shark measures were put in place. Helicopters flew over Perth's beaches for four hours a day at a cost of $1m (£670,000) a week, and beach patrols were extended from 6.30am until 7.30pm. Swimmers were urged to swim only off the patrolled beaches. Tourism Council Western Australia chief executive Evan Hall said he feared that the area would no longer be known for its beautiful beaches but instead for dangerous shark attacks: 'Without a doubt it could potentially have an impact on our image as a safe destination. Image can be everything, particularly in a tight market like tourism, so we've got to do everything to reinforce that we are a safe destination and that is well understood.'

- On 27 November 2010, British plasterer Michael Utley, 30, disappeared off Native Dog Beach after a Great White was seen in the waters. Sergeant Bob Scott from the Jerramungup police station said he first became aware of the missing swimmer at 4pm on the day Mr Utley went missing: 'It's a pretty rough coastline; the conditions were atrocious where they went in to have a swim and where he was actually swimming, there was a lot of foamy water about, visibility was poor because of that; generally the water is very clear. Unfortunately we are in a position where we are conducting recovery enquiries because there is no chance that he has survived.'

A massive search was launched, including a police helicopter flying over the beach, which lasted several days. Said Sgt. Scott: 'There was a shark between 2 and 3m [6.5 and 9ft] long, but we weren't able to identify the shark as it was quite dark. It was more of a distraction than anything else; it hasn't hampered the search. The helicopter did go closer to investigate the shark and went down into deeper water. However, it was seen further along an area in Dylan Bay, which is where our search was concentrated much closer to the shore.' Utley's body was eventually found at neighbouring Dillon Beach.

- On 17 August 2010, a mother's premonition came horrifically true when one last surf before flying off to a new job ended in tragedy for 31-year-old Nicholas Edwards. He was fatally attacked by a shark while surfing near Gracetown, off Western Australia's south-west coast. One witness told how Edwards lost his board and did not surface again. A rescue crew was launched and the surfer was eventually found unconscious on rocks near South Point with a huge gash in his leg. Brave efforts were made to resuscitate him and a tourniquet made using a surfboard leg rope but there was a 20-minute wait for an ambulance and Edwards died on the way to hospital.

One rescuer, Eddie Kilgallon, recalled: 'We heard the shout "Shark!" Shark!" There were about four of us in the water and we all freaked out. We started paddling for shore and a pod of seals came up in the middle of us. I thought it was a shark at first and thought, "This is it. I'm gone, too." When we got to Nicholas, I held what was left of his leg together. The boys were giving him CPR, just talking to him and doing everything they could. We just kept on encouraging him. We noticed that he had a wedding ring on and we constantly reminded him of the people that loved him and tried to pass that love onto him. His face was white but we started to get the colour back into him; that's when I thought he might have a chance. But the back half of his leg was totally chomped – it looked like one big shark bite.'

Local police sergeant Craig Anderson praised the rescue attempt, saying an 'outstanding' effort had been made to try and save Edwards, who was due to fly out to Goldfields to work in the mining industry and had only recently arrived in the area with his wife and two young children.

Edwards' mother, Leona Lindner, later spoke from her Adelaide home about the fears that had always haunted her. She said: 'I was worried, always worried, and lived in fear that I would get this phone call because it's dangerous, absolutely dangerous.'

Meanwhile, Eddie Kilgallon said the small community in Gracetown was also coming to terms with the death: 'It's a time thing, the day after. It's really hard to know where anybody's mindset's at.' Local councillor Ian Earle also spoke of the community's shock – but added that it would not deter people from going into the water.

Another rescuer, Bob Alder, said it was 'an honour to try to save a "brother". We treated him real good, we did everything we possibly could – he didn't die alone. If I had been attacked, that's how I would like to be looked after. We're surfers – it could happen to any of us at any time.'

After Edwards' death, his mother told how he had previously had an encounter with a shark when he lived on the Gold Coast: 'He was surfing up there one day and he had a very near miss with a shark up there and he paddled in like mad, and he was really shook up and he said he was scared. I said, "Don't go in there ever again," but no, Nick had to go.'

On 21 August, over 100 surfers paddled out and formed a heart shape of the beach where Nicholas died. Among the mourners were his wife Melissa and their two children, Lucy, 2, and Nathan, 10.

On 20 December 2008, diver Brian Guest (51) was killed by a Great White shark as he swam at Port Kennedy beach, south of Perth. Witnesses reported seeing a dorsal fin and thrashing in the water before the sea turned red and Guest vanished. The father-of-three, who had been an active campaigner for the protection of sharks, had been looking for crabs with his 24-year-old son Daniel when he was attacked about 30m (98ft) from the shore.

'I just saw a big splash and then the shark rolled over in the water with the guy and then there was no body or anything,' said one witness. Guest's shredded wetsuit was later found but his body was never recovered. Aerial searches were carried out and a 5m (16ft) Great White shark was spotted swimming in the area. Beaches were closed after the attack.

Brian Guest had penned a somewhat prophetic posting on the *Western Angler* website forum in 2004: 'I have always had an understanding with my wife that if a shark or ocean accident caused my death then so be it, at least it was doing what I wanted. Every surfer, fisherman and diver has far more chance of being killed by bees, drunk drivers, teenage car thieves and lightning. Every death is a tragedy – regardless of the cause – but we have no greater claim to use of this earth than any of the other creatures we share it with.'

On 8 April 2008, 16-year-old surfer Peter Edmonds died after being attacked by a shark at Ballina, on the New South Wales north coast. He suffered extensive bites to his left leg and body. Edmonds was bodyboarding with his friend Brock Curtis off North Wall beach at around 8.30 in the morning when the attack happened. Brock described how he saw a 'big, grey shadow' go by, realised Peter was in danger and paddled to his rescue. He added: 'In the water I was in line with him and noticed that he was in a bit of trouble. As I headed towards him it looked like he was catching a wave and was heading back to shore.

Then I saw him on his back, with his head above the water, so then he turned so he was face down. I thought he was only joking, so I went over to him and as I flipped him over, I saw his leg. I tried to resuscitate him but he didn't make one noise.'

A spokesman for the area Surf Life Saving group, Craig Roberts, said Edmonds suffered two large bites: one to his leg and the other to his body. 'The patient came in with another gentleman and lifeguards attended the scene. As you can imagine it was quite a distressful situation: he was unconscious and the lifeguards and ambulance officers had some severe haemorrhage to deal with. Certainly it's one of those things that you're trained for but hopefully it never happens as well. The lifeguards did an excellent job and worked with the ambulance and the police service, but obviously it is a traumatic experience.' Roberts added it was not common for sharks to be present off North Wall beach at that time of the year.

One local said fishermen had earlier seen four or five bull sharks close to the attack site. All beaches were closed. Peter was only 50m (164ft) from the shore when he was attacked; the beach area is normally patrolled, but crews had not yet arrived that day. Surfers' websites say North Wall beach, just north of the Richmond River estuary, is popular but warn of the danger from sharks. Of course none of this was any comfort to the victim's family. His father Neil saw the injuries and commented: 'We saw the bite on his leg and you wouldn't think that sort of thing could happen.'

- On 7 January 2006 church group member Sarah Kate Whiley, 21, was attacked and killed by up to three sharks as she swam in chest-deep water at Amity Point, off Queensland's North Stradbroke Island – an area protected by inflatable drum lines to keep sharks at bay. Witnesses at first thought her screams of 'Shark!' were said as a joke. They were horribly wrong as Whiley had both her arms torn off and her body and legs savagely mauled. Josiah Topou was sitting on the beach with his wife and five children when he heard the screams. He later told of his attempt to save the victim – even though the sharks were still close by: 'As I got to her and gathered her up, she looked at me and said just two words: "Help me." As I was going out, I knew there was danger; I was bringing her in. There were actually dolphins between me and Sarah and the sharks were on the other side. She was bleeding badly and telling people to get away. They had to hold her down, did the best they could to comfort her.'

  Whiley was taken by helicopter to Brisbane's Princess Alexandra Hospital, where she died from massive blood loss and shock. Paramedic Lachlan Parker said she was barely alive when they reached her: 'She had lost significant amounts of blood. The patient had what we call altered level of consciousness where we weren't able to communicate with her.' A police spokesman said the young woman 'went down under the water. After about five or six seconds the deceased came out of the water and screamed "Shark!" and of course people at the time thought she was joking – until they saw the blood. She was bleeding quite heavily. We're of the opinion from what we have seen and been told that there was more than one shark involved – it could have been up to three.' Police divers and fishermen tried to hunt down the sharks to 'retrieve what we can, but realistically it's virtually impossible.'

Beaches were closed after the attack while the search took place. Whiley's death was the first at a beach protected by Queensland's Shark Protection Programme since it was put in place some years earlier. It was believed that Bull sharks were the attackers and may have mistaken the young woman for prey in waters made murky by the previous night's storm. There was also some confusion over whether Whiley was swimming with a pet dog at the time of the attack. South Australian-based shark expert Andrew Fox said: 'It may mean that Bull sharks have moved into the area and are feeding. They are a pretty large, robust shark.' One witness added: 'I am sure she was swimming with her dog because I saw it come flying up the road, all wet and shivering and whimpering. Then a little boy came running up and said the girl had lost her leg and arm, and everyone ran out of the house towards the beach. It was just a little black-and-white dog, but he was crazy so I locked him under the house.'

And it seems a fatal shark attack was anticipated by the locals. Amity Point fisherman Miles Scott said:

'We've been waiting for this for a long time; we've always thought someone was going to be taken here. It's nothing to see ten or a dozen Bull sharks under our boat when we are crabbing and they are really aggressive – they are not like normal sharks.' Long-term resident Brad Ross added: 'The shore just falls away into 30m (98ft) of water and there are plenty of Bull sharks there. People know when they enter the water they are stepping into a shark habitat.' On the morning of the attack, the drum lines had been re-baited (the last shark attack on North Stradbroke Island had taken place in 1973).

Josiah Topou, who risked his own life to enter the waters in an attempt to save Whiley, was later granted a bravery award. About three months after the fatal attack, Topou's sixth child – a daughter – was born. Her name was an obvious choice. 'We though God's taken away one Sarah. My wife was six months pregnant at the time so we thought we'd name our daughter Sarah in memory of the one God took away,' he said.

- On 24 August 2005, Jarrod Stehbens (23) was killed off Glenelg, South Australia as he helped Adelaide University marine biology students collect cuttlefish eggs for research. The attack happened as he was reaching the surface of the water and he died of massive injuries. Two other divers tried to fend off the shark. Stehbens was an honours graduate in marine biology and had planned to complete a PhD in Germany. His father, David, said he 'was doing exactly what he wanted to be doing when it happened. He loved the sea. He was a very experienced diver, with probably over 190 dives, and he knew what it was all about.' Mr Stehbens said Jarrod would not have wanted the shark involved in the attack to be killed. Meanwhile staff at the university discussed whether to suspend its diving programme.

Jarrod's air tank and buoyancy vest were later recovered near the scene of the attack. Again, the culprit was believed to be a Great White. Shark expert Andrew Fox said: 'The Great White is really the only large predatory shark that's capable of actually taking a diver.' His father Rodney – himself a survivor of an attack from a Great White – said the Spencer Gulf area was perhaps the best feeding ground for Great Whites in the southern ocean, adding: 'It's probably the best restaurant in the southern ocean.'

- On 19 March 2005, 26-year-old Geoffrey Brazier died from 'massive injuries' as he snorkelled with a group of tourists on Australia's Coral Coast. He was savaged by a huge shark and died instantly. Witness Rod Plug said Brazier stood no chance as the Great White lunged towards him: 'It was very large, very wide – we estimated over 6m [19ft] long. Very dark on top, white underneath, a pointed nose and a dorsal fin that was almost a metre (3ft) high.'

  Brazier was a deckhand on a charter vessel, the *Matrix*, which was on its maiden voyage from Perth to Kimberley when the tragedy occurred. It was moored at Wreck Point on Pelseart Island, part of the Abrolhos Islands group and 60km (37 miles) west from the port of Geraldton. A police spokesman said: 'A 26-year-old deckhand was in the sea snorkelling with passengers of the vessel when he was seen to be taken by a large shark described as being approximately 6m (19ft) in length. It would appear that he died instantly. His body has not been recovered.'

The Western Australia Fisheries Department, police and State Emergency Service officers launched a search for the shark, which they intended to kill and use for forensic investigations. Fisheries officer Rory McAuley said there were two reasons why they were hunting the shark: 'One of the reasons, I believe, is that they're looking for the remains of the victim, which will obviously be necessary for a coronial inquiry. I guess the other main reason is to ensure that the shark poses no ongoing risk to public safety.' Meanwhile, rock lobster fishermen and their families on the islands were urged to stay out of the water while authorities searched for the shark. 'We have gone through the fisheries and people who are on the Abrolhos Islands and warned them in relation to a person being taken by a shark. Obviously the message would be at this stage for all persons to remain in vessels and on land and not venture into the water,' said a police spokesman, adding: 'Also, the Abrolhos is a lot of islands and has a big volume of fish and many, many sharks. We don't think there's anything left.'

Greg Davis from the Geraldton Professional Fishermen's Association said those who live and work at the Abrolhos Islands, especially surfers and scuba divers, were in shock: 'It's always something that's in the back of your mind but some people have said that they won't go surfing at the Abrolhos again, stuff like this. I've been going to the islands since I was born – 42 years – and it's the first time there's been a fatal shark attack as far as I'm aware in that time.' But he added that sharks were not uncommon in the area: 'You see a shark or two but it's par for the course because they live out there and the islands are not far from the edge of the shelf. It's quite deep water and that's their habitat.'

- On 16 December 2004, 18-year-old surfer Nick Peterson was killed off an Adelaide beach in South Australia by two sharks. Adam Floreani, one of three 16-year-old friends in the boat towing Peterson on his surfboard, described the attack: 'It got his left arm and took him around the boat and then another shark came in and they just took him to pieces.' Another said: 'It came up from nowhere – he didn't see it for a second before it happened. He went down fighting, he didn't give up.' Police said the attack had happened when Peterson fell off his board: 'Apparently one shark took him and the other came in and took the rest.'

South Australia A Sea Rescue Squadron spokesman Frasier Bell said one of the Great Whites was believed to be about 5m (16ft) long and the other about 4.5m (14ft): 'Mr Peterson fell off the surfboard and the shark appeared and took him; it tore him apart. They were just boys having a good time – the weather was perfect and they were just doing what young lads do. They tried everything they could think of to save him but unfortunately the sharks had taken him by that stage. They're in deep shock, they're wrecks – you can just imagine what the victim's parents are going through.'

Despite the horror of his son's death, Philip Peterson insisted he did not want a shark hunt. Visiting the beach with his wife Leonie, Peterson said: 'We acknowledge that the sea is, in fact, the shark's domain. We don't, and I certainly don't, advocate the indiscriminate killing of any shark – they are to be admired, appreciated and respected, and Nick knew that.'

It was believed the killer sharks were the protected Great Whites species but after a meeting of government, police, fisheries and coastguard officials came the order that the sharks be found and destroyed. 'Any shark posing a threat or risk to human life should be destroyed, that is unequivocal,' said acting South Australia Premier Kevin Foley. He also defended no action being taken following a shark sighting the week before – even though it was probably one of those that took Nick Peterson: 'Should action have been taken earlier? Every summer we are confronted with that dilemma. We don't know at all whether the shark that took this poor victim was the shark that was sighted last week. What we don't want is a standard culling approach to Great White sharks at the beginning of every summer.'

Several sightings of the killer sharks were made near West Beach as about 50 emergency service boats supported by helicopters searched for the animals and any remains of Nick Peterson. The beaches remained open but with a police warning: 'People who use our beaches need to consider the risk of shark attack'. Two months after the fatal attack, a new, seven-day helicopter service was launched both to search for sharks and to rescue anyone from the water. Before, the patrols had only been made at weekends and Nick Peterson had died on a weekday. Philip Peterson described the new helicopter as 'Nick's gift,' adding: 'The water should never run red again.' The helicopter came mainly from private funding.

Some time later South Australia's top 46 surfers took to the water to take part in the first-ever Shark Shield Nick Peterson Classic. The event included a ring of 20 surfers in the waves in a gesture of remembrance and raised $3,000 for the Nick Peterson and Surf Life Saving fund.

- In July 2004, surfer Brad Smith (29) was killed by a Great White. He had been with four other people at Left Handers beach, south of Gracetown, near Margaret River in Western Australia when he was knocked off his board and mauled. One of the two teenagers who pulled Smith from the water said the shark was 'as big as a car.' Witnesses said he had 'fought for his life' with one adding: 'When we looked out there was this guy getting eaten by this massive shark and there was another shark circling around him.' Police spokesman Peter Reeves said several teenagers had come forward to give statements, describing how the attack lasted about a minute. They had dragged Smith's body back to shore, hoping they would be able to save him but he was already dead from blood loss and horrific injuries. Senior Constable Reeves added: 'Obviously what they have witnessed is very graphic and very traumatic and everyone handles things differently but yes, some of them are suffering and are in shock.'

Authorities closed the beach after the attack and erected shark warning signs; a helicopter and boat search was launched to track down the shark. Fisheries officer Tony Cappelluti said officers were authorised to kill the shark that attacked Smith if they were able to locate it but he said this was a last resort: 'We are doing all we can to find it; the boat is out there and we have a chopper up in the air but the search conditions are pretty poor. Obviously we would want to herd it offshore to deeper water so that it no longer remains a threat to the public but we do have ultimate authority to catch it. If they're going to become a danger to the public, or if we believe they already have been responsible for a fatal or serious attack, then I think the community would expect us to try to alleviate that risk. But Brad Smith's brother, Stephen, said he had died doing what he loved best and killing the shark would be 'senseless revenge.'

- On 30 April 2002, Paul Buckland, 23, a professional shell fisherman from Port Lincoln in Western Australia, was bitten and fatally injured by a 6m (19ft) long female shark. Even the shark repellent device he was wearing at the time did not save him and he died in his diving partner's arms. Shannon Jenzen, 24, told how he had pulled his friend from the water before Buckland succumbed to his injuries. Jenzen described how they had set off early that morning and he had made his last dive for scallops at 11.30am before swapping with Buckland, handing over the safety anti-shark pod to him. About ten minutes later he heard Buckland shout out: 'I heard him call out, and I knew exactly what he'd seen and I turned on the motor of the boat. I didn't see the shark at the time but I went out to him. There it was – it was huge.'

In a desperate attempt to save his partner, Jenzen hit the shark – described as being as big as the boat he was in – with the side of the boat: 'It didn't move at all; it ignored me. I saw the shark attack, but it never took him down. I finally got Paul into the boat and could feel the pulses sent out by the Shark Pod. I turned it off and got it off him as quickly as I could.' He wrapped his friend in tarpaulin used to protect scallops from the sun. Paul Buckland had lost his left leg and a greater part of his torso. Jenzen tried to call for help on a mobile phone but the batteries were flat; the on-board radio was not working either. He said that he eventually 'steamed' towards a group of fishermen a short distance away to get help.

The fatality appeared to have come about because Buckland had not turned on the anti-shark device when he was handed it, but only activated it when he saw the shark approaching. At an inquest in March 2003, another scallop diver – Russell Barber – told how he had witnessed the attack from a boat: 'He had his shark pod on and it was working because he was giving his diving partner shocks as he pulled him into their boat. He was dead within seconds – he had no leg and no hip area left.'

Local fishermen demanded the shark – believed to be a Great White – be hunted down and killed to prevent further attacks. One said: 'The fearless monster made herself at home in this area, which strongly reminds one of scenes from the film, *The White Shark*.' But in a country where Great Whites enjoy strict protection, this was not an easy option. Will Zacharin, director of fishery matters of the South Australian Fishery Ministry, said he wanted to monitor the shark's whereabouts and if necessary, simply drive it away from the area. 'This shark will only be killed if we have no other alternative because these sharks are protected in Australia,' he said. Experts said killing the shark would have no positive effects on safety in the region because several white sharks lived in the waters and to single one out and remove it would not make the region any safer for commercial or leisure-time divers.

Production of the Shark Pod – used by the army and competitors in the Sydney Olympics, and which produces an electrical field around a diver –

ceased when a new shark defence device, the Shark Shield, was launched. But there were those who still maintain the Pod should have saved Buckland, with one spokesman saying: 'Thousands of people have used this technology and frankly, the limited facts about this incident don't add up.' He said there had been no previous reports of anyone being attacked while wearing the Pod.

Fisherman Kang Suk Lee, 52, accidentally drowned while 42km (26 miles) out at sea, off Lake Macquarie, New South Wales, on 2 April 2002. His remains were found in a 3m (9ft) Tiger shark on 21 April 2001. On 9 June, Arthur Applet, 75, was fatally attacked, reportedly while hiking on the edge of the beach at Lord Howe Island, New South Wales. His remains were later recovered from the inside of a Tiger shark.

- On 6 November 2000 Ken Crew, 49, was killed at North Cottesloe beach, Western Australia. His leg was ripped off despite the efforts of two men to fight off the shark – believed to be a Great White. Crew was returning from his daily swim when he was attacked about 30m (98ft) off the beach. Fellow swimmer Dirk Avery, 52, suffered deep wounds to his feet as he tried to fend off the shark (the second rescuer was unhurt). The men were part of a group of about ten swimmers returning to shore when the shark attacked in waist-deep water in front of scores of witnesses at the popular beach. Among them was owner of the Blue Duck Cafe Kim Gamble, whose establishment overlooked the scene of the attack. He described how the water turned red: 'From the balcony I could see this huge shark – it was really huge. There was a whole sea of blood and it was pulling the person. It's something I wouldn't want to see again.'

Father Brian Morrison, a Perth priest and friend of the dead man, also watched in horror from the cafe: 'Two men went to rescue – one was hurt on the leg. He's in hospital, but we don't know if it's too serious. The other man, I think, is the hero because he really tried to drag him away from the shark. He seemed to run in and tried to get rid of the shark in some way.' Morrison administered the Last Rites to Crew as he lay on the beach.

Family friend Lee Tate said Dirk Avery was 'very brave' to have tried to kick the shark away from Crew: 'When Ken was attacked by the shark the other man was attempting to kick that shark away, standing on the reef. The shark was so big and so many people saw it and had called out "Shark!" but by that stage the attack had happened and people were basically helpless to do anything about it. Mr Crew was helped onto shore with a severe loss of blood and unfortunately, as I understand it, he died on the beach.'

Tate said Crew's wife Robin was walking on the beach when the attack happened and although she did not see the incident, she was at the scene soon after her husband was dragged to shore. According to him, she had suffered 'severe shock and trauma.'

Beaches from Fremantle to Hillary's remained

closed as police and fisheries officials mounted an intensive sea and aerial search for the shark. A 4m (13ft) Great White was spotted at Cottesloe around three hours after the attack and was followed for about 45 minutes by officials in boats and helicopters before they lost sight of it. Fisheries Western Australia research director Dr. James Penn emphasised the Great White was a protected species and would only be killed as a last resort and after special permission had been obtained from environmental agencies.

- On 24 September 2000, New Zealander Cameron Bayes, 25, was dragged off his surfboard at Cactus Beach on South Australia's west coast. His attacker was a Great White and Bayes sustained horrific fatal injuries in the 90-second horror. Though his surfboard was recovered, his body was never found. Witnesses said the shark was about 4.7m (15ft) long and thrashed about in circles after grabbing Bayes. They were surprised at the 'severity' of the attack and that it had happened so close to shore – about 50m (164ft) from the beach. Said one: 'A really large shark just came in and attacked him, thrashing around in a circular motion around him. He seemed to get back on his board, and was paddling back and it came in again. It was just horrendous – it totally took him out. It just seemed to roll on its belly then it thrashed around a bit more, then it seemed to release the surfboard but there was nothing left of the guy.'

  Added another: 'It didn't attack him from above, it just created a whirlpool which dragged him down.

I just saw a flash of red as the wave came over and then it was all over. We saw the last 30 seconds and that took forever; we were in disbelief that it could be that close. There was certainly a lot of force in it and it was very savage. The thing really wanted him and it wasn't going to let him go.' Another witness, Jeffrey Hunter, would later give a full and graphic statement at the inquest into Bayes' death:

I looked at this man and saw a shark coming out of the water around him; it was thrashing on and around the man. Its head was shaking from side to side and thrashing in a circular motion; there was a bit of blood in the water. I could see the man and the shark at the same time and then they both went under the water. Where it was happening it just looked like it was all shark – I couldn't even see the surfboard. The man came up, I only saw his head and body; he seemed to be back on his board and started paddling. He made about 3m [9ft] and then the shark went for him again. The shark was thrashing and shaking its head; there was a lot of blood. The shark seemed to take him underwater and there were only a couple of bits of board left. It took him under and there was a bit more thrashing under the water.

A couple of minutes later I saw it surface about 500m [1,640ft] out; it thrashed again and released the main bit of the board. It rolled over and I could see the underside of the shark: it was clearly white; the top part of the shark was a dark grey. The main part of the board floated towards Caves.

Conditions at the time were overcast – no wind, misty,

with a 3–4ft swell. I know from surfing this beach that the water where the attack occurred is about 2m deep; I would estimate the shark was between 4 and 5m long.

Ron Gates, owner of a nearby campsite, said after the attack he was 'shocked, horrified and sick,' adding: 'This is the fourth attack since I've been here so it's not the first, but it's the first fatal so it makes a big difference. Everyone is treating this differently to previous attacks; everybody is totally devastated.' Gates immediately put up signs closing the beach while female campers comforted Bayes' distraught wife until an ambulance arrived (the couple had been on an extended honeymoon, travelling around Australia).

There have been near misses throughout Australia, too and although escaping death, some of the victims were left with enduring injuries:

- On 17 March 2011, Lisa Mondy was grabbed first by her face and then her arm while wakeboarding near Nelson Bay on the New South Wales coast. Her injuries were horrendous and it was at first thought she might lose the arm. She said: 'My face was like minced meat but it wasn't its fault, it was just one of those things. I didn't see it coming. The first I knew was when it grabbed my face. As weird as it sounds, the first thing I thought was, "Hey, I've always stuck up for sharks, why are you doing this to me?"'

  She was left with horrific scarring, a disfigured mouth and neck and little movement in her arm: 'It's pretty awful under there but at least I've got it,' she said. The brave woman was visited by another shark attack survivor, David Pearson, 48, who was attacked a week later at Crowdy Head – and who happened to be receiving treatment for a mauled arm at the same hospital. Mondy was also contacted by Paul de Gelder, a navy diver who lost his right hand and part of his right leg during a shark attack of February 2009, which took place near Garden Island, Sydney Harbour.

- The day of 30 October 2010 saw one of the most miraculous shark stories ever when a young woman was saved by a quick-thinking swimmer, who grabbed an attacking shark by its tail! Elyse Frankcom was in charge of a group of tourists swimming with dolphins off Garden Island, about 10km (6 miles) from Perth on Australia's western coast. It was believed to be a Great White, which came up from the depths of the ocean and attacked Frankcom. Trevor Burns proved to be her guardian angel that day, after pulling at the shark's tail. One of the first rescuers on the scene, Fremantle Sea Rescue senior skipper Frank Pisani, said: 'He certainly was instrumental in making this a good outcome. As the shark bit the woman, it brushed aside Mr Burns, who grabbed hold of the tail of the shark, which then made it let go. The girl then started to sink to the bottom and he grabbed her and brought her to the surface and got her back on board the boat.'

Frankom sustained bites to both thighs and underwent six hours of surgery after being flown to hospital in a helicopter. But she was alive, thanks to the fast actions of her saviour. More than 30 people, including several children, were on board the boat at the time of the attack. Frankcom was working for Rockingham Wild Encounters, a company who organised boat and swimming trips. She was left with a solid reminder of her near-fatal encounter. Terry Houston from the company said: 'They pulled a shark tooth out of her butt, so Elyse is pretty proud of that – she's got that as a bit of a memento.' In fact, the young woman had to have 200 stitches in her wound.

Talking about her ordeal later, she said: 'I felt the force of being kind of pulled to the side and then I looked and saw a grey figure in front of me. I felt me inside its mouth – it was horrifying. There are a lot of people that don't make it and a lot of people don't survive.' Frankcom said she had only switched on her Shark Shield during the attack. Burns said 'there was no way' he was going to let the shark take the young victim and admitted she probably would not have survived, had he had not come to her aid. Frankcom later got to thank her rescuer, calling him her hero. Shortly before the attack she had posted a message on her Facebook page saying: 'If I get attacked or die, at least I die happy and doing the thing I love.'

- On 13 February 2010, Australian grandmother Paddy Trumbull, 60, survived a ferocious mauling as she snorkelled near the Whitsunday Islands, Queensland. Doctors said she was lucky to be alive. Trumbell was snorkelling from a chartered boat when she felt 'the most almighty huge tug.' Describing the attack, she said: 'I turned around and I saw this huge shark. I then thought, "This shark's not going to get the better of me" and I started punching it on the nose – punching, punching, punching. And then it got me under the water, but not much because I started kicking at its neck. I had a bit of a tug of war with it. He was huge, at least 2m [6.5ft], but his nose was so hard. I could see the teeth glaring, but I wasn't seriously panicking – all I was thinking about was surviving.'

Trumbull was pulled back on to the boat and given first aid before being airlifted to Mackay Base Hospital, where she underwent surgery for severe injuries to her buttocks. Surgeon Paul Flanagan said: 'We can estimate that she lost about 40 per cent of her blood volume from the degree of shock that she had when she came in and the fact that we required to give her several units of blood. We were staggered when we saw the size of the wound.' Trumbull said she was happy to be alive: 'I think they're going to get me a counsellor to sort of sort it out and I have to have a new, remodelled bottom, so that's a positive.' There had been no warnings of possible shark danger as an attack had not taken place in the area for 13 years.

- On 11 January 2009, there were three Australian shark attacks in 24 hours. The first was at Fingal, off the northern coast of New South Wales. Jonathon Beard, 31, was bitten on the leg while surfing but managed to swim ashore. He was left with a large bite hole in his left thigh from what was believed to be a Great White: 'I didn't see a thing. It lifted me out of the water and started shaking me, like I was riding a bull or something. Then I felt a burning sensation and the sea just turned red – it was like a *Jaws* movie. I don't think I have ever paddled as fast as I did then, but somehow I knew I was going to be alright. I knew it hadn't hit an artery.'

In fact Beard had had a very lucky escape indeed. Donald Pitchford, Gold Coast Hospital's executive director of surgery and critical care, said the shark had come within about 2cm (0.6in) of a femoral (groin) artery: 'If that had happened then by the time he got back to the beach, he would have bled out.' Despite the encounter Beard said he had no plans to give up surfing, but he was apprehensive about returning to the scene of the attack: 'We're going to get a carton and sit on the beach, but I don't think I will go surfing there again. I agree that perhaps rogue sharks should be killed, but I don't think we should go around killing them all – I was in his territory.'

A few hours later, off a beach on Tasmania's east coast, surfer Hannah Mighall, 13, was attacked by what was believed to be a 5m (16ft) long Great White. Her cousin, Syb Mundy, was hailed a hero for saving her. He said: 'She just flew up in the air and got dragged under – the shark had given her a nudge and she disappeared. She came back up and went down again a few times and I saw the shark come up out of the water. It was thrashing her around like a rag doll, but she kept her head together. I can remember seeing the shark's eye come out of the water and the head and I was going to try to poke it in the eye, if I could get close enough. The shark started circling us and coming up underneath us and when it did that, we stopped and turned to face it so we could push it out of the way. Luckily a wave came along and she was on her back, and I said, "Hannah, no matter how weak you are, try to hang on – this wave is going to save our lives." But the shark actually got on to the wave. This thing started surfing towards us and we just headed straight to the beach. I don't think we had anything to do with it – we were just powerless against it – it was such an intense creature. I think it just didn't like the taste of her.'

After releasing Hannah's leg, the shark grabbed her surfboard (attached to her ankle by a leash). She was taken to hospital with leg wounds. Her father, Malcolm, said: 'She's been in pretty good spirits but she's had trouble sleeping. Every time she shuts her eyes, she sees you know what.'

Not long after this incident, despite suffering 40 puncture wounds to his leg, snorkeller Steven Fogarty, 24, survived an attack at the mouth of the Illawarra River, near Wollongong, south of Sydney. He was in water just 2m (6.5ft) deep when he felt the shark nudge his calf and fought it off by punching it in the face. He said: 'Something just grabbed me from behind. At first I thought someone was playing a joke but then I saw the blood all over both feet and had a quick look to make sure both legs were there, and they were there. I started swinging punches at it. I think I got one in – that's probably what got rid of him. All I could see was a white cloud with bubbles around it. I started screaming, got on my back and started back-pedalling.

'I was in the water a good couple of minutes. A couple of people drove past in a boat. One bloke just went straight past me, just looking at me. I yelled "Help! Help! Help!" Bleeding heavily from his injuries, Fogarty eventually managed to flag down a boat. He was treated by paramedics before being taken to Shellharbour Hospital. District manager for Illawarra Ambulance Service, Terry Morrow, described the injuries: 'There were approximately 30–40 teeth marks around his right calf that appeared to be from the jaws of a shark. They were clean, simple cuts, like little razor blade cuts.'

After the ordeal, Fogarty had some advice for water users: 'Don't swim on your own – I've been told a thousand times!'

Grandmother Joanne Lucas, 50, was hailed a heroine after risking her life to rescue a swimmer being savaged by a Great White on 11 May 2008. With no thought for her own safety, Mrs Lucas braved the waters, crossing nearly 100 yards to pull schoolteacher Jason Cull, 37, back to shore at Middleton Beach, near the south-west Australian town of Albany. Cull had been swimming with dolphins when the 5m- (16ft) shark tore two chunks from his leg to leave him screaming in agony. Those screams reached the ears of Lucas, who tells an amazingly heroic tale: 'Instinct just kicked in. I didn't even have to time to think about it, which is truly amazing. I just saw someone thrashing in the water and crying "Help me! Help me!" At first I thought there was just a dolphin in the water, but someone else was screaming that he'd been attacked. I just thought I had to get in there. I got to him and he said, "Thank God! Thank you so much – a shark has attacked my leg." I kept thinking that I've got to get him in before it turns around and comes after us. I was thinking, "I have to beat the shark to the shore." He had huge chunks taken out of his leg, his calf and the knee area – I kept talking to him because I didn't want him to slip into unconsciousness.'

Mrs Lucas then ran to fetch towels to wrap around Cull's wounds and sat with him while they waited for paramedics to arrive. He was rushed to Albany Regional Hospital to undergo surgery. Talking from his hospital bed, Cull said: 'It all happened so quickly – I just saw this grey shape coming towards me. Initially, I thought it was a dolphin. When it came up and banged straight into me, I knew it was a shark. I was more concerned about getting out of its mouth because it was dragging me backwards underwater. I just remember being dragged along backwards. I was trying to feel for its gills but I found its eye and I stuck my finger in and that's when it let go.'

Police Sergeant Roger Creamer described Lucas's actions as 'extremely brave.' Local lifesaver Tom Marron echoed this praise, saying: 'She heard him shout out for help and dived in with no regard for her own safety. He suffered a fair bite. If she hadn't followed her instinct, or had been a bit later, then the bloke could have bled to death or been dragged out by the shark. What she did was brilliant.' And Cull was fully aware of how his life had been saved by the fearless woman who went to his aid, saying: 'Thankfully she did – I do not think I would have made it the rest of the way.'

- On 20 April 2008, teenager Jamie Adlington was attacked by a Tiger shark at Crescent Head, New South Wales.
- On 23 January 2007, 41-year-old diver Eric Nerhus was almost swallowed alive by a 3.25m (10ft 6in) Great White shark off the fishing town of Eden, about 75km (46 miles) from Sydney. He was only saved by his lead-lined vest, which prevented him from being bitten in half. Nerhus had been with his son and other divers collecting abalone – and edible shellfish – when the shark suddenly attacked him about 8m (26ft) below the surface, grabbing him by the head. 'Half my body was in its mouth,' he later said. 'I felt down to the eye socket with my two fingers and poked them into the socket. The shark reacted by opening its mouth and I just tried to wriggle out. It was still trying to bite me. It crushed my goggles into my nose and they fell into its mouth.'

  Nerhus said that he managed to finally escape the shark's jaws after jabbing at its eye with a chisel he had used to chip abalone from rocks and which he was still holding despite the attack. But he estimated he had spent two minutes inside the shark's mouth, the mighty force crushing his facemask and breaking his nose. As he swam to the surface in a cloud of his own blood, Nerhus said he could still see the shark and feared it would attack again: 'It was just circling around my flippers, round and round in tight circles. The big round black eye, 5in [2cm] wide, was staring straight into my face with just not one hint of fear of any boat, or any human or any other animal in the sea.'

Fellow diver Dennis Luobikis, 53, said: 'The brunt of the shark's head and snout was taken by Eric's led weight vest. He said to me at the wharf that his vest saved him. Eric was actually bitten by the head downwards – the shark swallowed his head. But he is a tough boy, super-fit. He pushed the abalone chisel into its head while it was biting and it let him go and swam away. I would say that it would test anyone's resolve being a fish lunch.'

Nerhus was helped into his boat by his son Mark, 25, who witnessed the horrific attack – 'He came up to the surface and he was going, "Help, help, there's a shark! There's a shark!" I went over and there was a big pool of red blood, and I pulled him out of the water and he was saying, "Just get me to shore. Get me to shore."' A nearby boat radioed for help and Nerhus was airlifted to hospital, where he was treated for severe cuts to his head, torso and left arm. It is believed the shark that attacked Nerhus probably mistook him for a seal, which are common in waters off south-eastern Australia and attract sharks.

- On 4 September 2005, Jake Heron, 40, miraculously survived a horrific shark attack while surfing At Fishery Bay, near Port Lincoln, South Australia. The attack occurred on Father's Day as he went surfing not far from shore and his two children played on nearby rocks. The predator was a 4m (13ft) Great White. Heron's friend, Craig Materna (who was surfing close to him) watched as the Great White launched its attack and desperately paddled towards him: 'He was freaking out and yelling for help. No one saw the shark come up to him. It knocked him off the board; it pulled him under because the leg rope was attached to him. He kicked and punched the shark, I think, in the gill.'

  Despite Materna attempting to help his friend, he could only watch as it tore at the surfboard and then bit Heron through his wetsuit on his right arm and thigh. Heron then struggled to shore, yelling and holding his leg. Somehow he made it to his car, where towels were wrapped around his wounds. He was taken to the nearby town of Tulka, where waiting ambulance transported him to Port Lincoln Hospital. There, he had 20 stitches in his arm and 40 in his thigh. Materna observed: 'If the shark had bitten him in any part of his body, it could have been good night to him. Not too many people survive a shark attack but the temperature of the water might have helped slow down the bleeding and he didn't lose too much blood. Jake was lucky – the shark had a few goes at him; he got a shock. He realises he is lucky to be alive.' He added that although the two young children witnessed the attack on their father, they were 'just fine.' The local fisheries department launched a search for the Great White.

- On 22 October 2004, John Gresham, 59, was surfing with his son at Stockton Beach, New South Wales when a shark snapped at his board and sent him spinning into the air. Gresham escaped with only minor injuries to his foot. He said: 'I've never had nothing like this happen and I hope I never do again. I'll go back in the water, but I'll probably wait a couple of weeks.'
- On 25 September 2004, three men spearfishing and filming from an inflatable dinghy off Batt Reef, Queensland escaped with only a damaged boat when attacked by a 3m (9ft 10in) Tiger shark.
- In February 2004, Anthony Hayes survived an attack when his friend punched and poked a shark in the eyes until it let go. In March 2004, a Great White shark scared swimmers from the water at Cottesloe. Weeks earlier, professional diver Greg Pickering, 47, survived an attack from a Bronze Whaler off Cervante.
- On 30 January 2002, Andrew Cribb, 20, escaped with only minor bruises and cuts when attacked by a 3m (9ft 10in) Tiger shark while surfing near Fingal Spit, New South Wales.
- On 21 November 2000, professional diver George Lyons, 28, escaped injury when attacked by a 5m (16ft) Tiger shark at Orpheus Island, Queensland. Miraculously, the only part of him to come to any harm was his wetsuit and one of his fins, which were torn in the attack.

Queensland is one of the 'top ten' areas in danger of sharks. Since records began there have been 103 attacks and 38 fatalities. Some of Queensland's beaches are protected by drumlines or baited hooks aimed at catching sharks; there is also protective netting. The Queensland Government says: 'The nets are designed to catch sharks more than 6.6ft. (2m) in length, so that the more dangerous sharks aren't coming in close to shore.' Some environmental groups object to the nets, saying they are a danger to marine life and prevent free movement of sharks. In response to controversy over a baby Humpback whale killed in the nets in 2005, the Government released figures highlighting the success of the shark defence device – with 630 sharks caught in one year alone. Of these, 298 were bigger than 2m (6.6ft) and included a 5.2m (17ft) Tiger shark.

In the past 40 years there has been just one fatal shark attack at the 134 beaches protected by shark nets, but 15 fatalities in South Australia, 12 in Western Australia and seven in Victoria, at beaches with no nets. Wil Zacharin, executive director of fisheries of the Western Australian Department of Primary Industries and Resources, said the introduction of netting is 'not feasible' and 'there is more danger driving to the beach than on the beach.' Western Australian Fisheries Minister Kim Chance added: 'The general belief now is that nets not only prevent protection, they may attract sharks because fish get caught in the nets and sharks come in to eat them.'

Australia's shark attacks are rising: there were 20 in 2009, compared to 12 in 2008 and 13 in 2007. In November 2011, the West Australian Government announced new measures to 'educate' swimmers and divers about sharks and to reduce the number of Great White shark attacks. The plans would cost $14m (£9m) over the next five years and came after what was described as 'an unprecedented' number of shark attacks in West Australian waters over the last year or so. It was the most recent ones which had prompted the Government to establish a shark response unit to look into the possibility of shark nets, shark repellent devices and SMS warning systems, all aimed mainly at the Great Whites. Funding was also to go towards a tagging programme, research into shark behaviour and a review of local fish stocks. Fisheries Minister Norman Moore ruled out a shark cull to minimise threats to those in the water but said any research, 'will educate ocean-goers to make safer decisions.' He added: 'The whole intention here is to try and seriously understand the ocean and to give people the best advice we can possibly provide for them in respect of potential shark attacks.' A petition to oppose the culling of sharks was signed by 19,000 people.

AUSTRALIAN SHARK ATTACK FACTS

- Sharks live in all the coastal waters and estuarine habitats around the 35,000km (21,748 miles) of Australia's coast.
- Attacks are thought to be on the increase as the country's population grows and more people enter the waters for both recreational and commercial reasons.

- The Australian Shark Attack File lists 53 fatalities in the last 50 years, saying: 'Some years there are no fatalities recorded, other years there have been up to three in a year but the average remains around one per year. Yet each year thousands of swimmer-days take place on our beaches, harbours and rivers and the number is increasing with both increasing population and tourism.'
- The earliest-recorded fatal shark attack was in 1791, when a woman died off the north coast of New South Wales.
- In 1963, the last fatal attack in Sydney Harbour took place. The victim was a woman called Marcia Hathaway.
- There have been 298 shark attacks around New South Wales. Of these, 69 have been fatal, 167 non-fatal and in 62 cases the victims escaped without injury. The last fatality was at Ballina, Lighthouse Beach in 2008.
- There have been 301 shark attacks off Queensland. Of these, 89 were fatal, 186 non-fatal and in 26 cases, the victims escaped without injury. The last fatality was at Fantome Island in 2011.
- There have been 120 shark attacks in Western Australia. Of these, 19 were fatal, 80 non-fatal and in 80 cases the victims escaped without injury. The last fatality was at Rottness Island in 2011.
- There have been 62 attacks in Southern Australia. Of these, 19 were fatal, 35 non-fatal and in eight cases, the victims escaped without injury. The last fatality was at Coffin Bay in 2011.

- Compared to fatalities from any other forms of water-related activity, the number of fatal shark attacks is extremely low, with an average of 87 deaths per year from people drowning at the beach compared to the average of one person killed by shark attack annually over the last 20 years.

The Australian Shark Attack File produced precautions against shark attacks:

1. Swim at beaches that are patrolled by Surf Life Savers.
2. Do not swim, dive or surf where dangerous sharks are known to congregate.
3. Always swim, dive or surf with other people, preferably at patrolled beaches.
4. Do not swim in dirty or turbid water.
5. Do not swim while bleeding.
6. Avoid swimming well offshore, near deep channels, at river mouths or along drop-offs to deeper water.
7. If schooling fish start to behave erratically or congregate in large numbers, leave the water.
8. Do not swim with pets and domestic animals.
9. Look carefully before jumping into the water from a boat or wharf.
10. Do not wear jewellery or shiny objects as the reflections could be mistaken for those from fish scales.
11. Do not swim at dusk or at night.
12. Do not swim near people fishing or spear fishing.
13. Do not swim near fur seal colonies, especially during the pupping season.

14. If a shark is sighted in the area, leave the water as quickly and calmly as possible.

The Australian Shark Attack File is coordinated at Sydney's Taronga Zoo and is associated with the International Shark Attack File (managed by the American Elasmobranch Society). The aims and objectives of the Shark Attack File are to:

- Chronicle all known information on shark attacks from Australian waters past and present and to record future attacks.
- Provide source material for scientific study to identify the common factors relating to the causes of attacks on humans.
- Provide summary information for public education and awareness and/or publication by the media.
- Publish information resulting from analysis of the acquired data.

CRITERIA FOR INCLUSION

Any human/shark interaction:

- Where there is a determined attempt by a shark to attack a person;
- Where injury occurs by a shark during an attempt to attack a person;
- Where imminent contact was averted by diversionary action by the victim or others (and no injury to the human occurs);
- Where the person is alive and in the water at the time of the incident;
- Where the equipment worn or held by the person is bitten or damaged by the shark during the incident, and;

- Where there is a determined attempt by the shark to attack a kayak, surfboard or small dinghy operated by a person.

As part of a worldwide study into shark behaviour, data from the Australian Shark Attack Files helps to identify the existence – or absence – of common factors relating to the cause of attacks on humans. The research project is conducted in three stages:

1. Compile information on each recorded attack in Australian waters.
2. Assimilate, categorise and transcribe data to computer.
3. Analyse acquired data and publish results.

However, the group says a lot more research is needed: 'This project is aimed at understanding and documenting the behaviour of sharks when they interact with humans. This information will contribute to conservation of the species and their environment through education and specific research projects. There is a need to learn more about the shark's normal behaviour, as well as in circumstances of human interaction.'

AUSTRALIA'S SHARK SPECIES

There are at least 166 different species of shark in Australian waters. These include:

- Great White
- Tiger shark
- Bull shark
- Hammerhead

- Blue shark
- Mako
- Whale shark
- Dwarf Sawfish
- Queensland Sawfish
- Freshwater Sawfish
- Green Sawfish
- Dindagubba
- Narrowsnout Sawfish
- Grey Nurse shark (critically endangered)
- Speartooth shark (critically endangered)
- Northern River shark (endangered)

CHAPTER 5

NEW ZEALAND'S NASTY SHARKS

'I WAS CONCENTRATING ON THE SHORE AND GOING AS FAST AS I COULD…'

On 16 April 2011, Laine Hobson was attacked by a Bronze Whaler shark while surfing at Snapper Point. He wrestled with the shark and his hand was bitten. Hobson first saw the predator's snout, then felt a nudge and became terrified as the shark leapt onto his surfboard: 'It wasn't like *Jaws* coming at me viciously; it came up and nudged me on the thigh while I was sitting on my board. It gave me a shock but I pushed it away with my hand, then it cruised around in front of me and gave me another nudge on the other leg. I was then trying to keep the board between me and the shark – I was probably screaming by then – there was another guy close to me. Then the shark nudged me off my board and leapt on top of it and was thrashing about. Everyone was out of the water at that point. It was quite horrible actually because this wave came along and picked up everyone else and carried them in, but it missed me. I couldn't get a break.

'I don't know if the shark followed me in. I was concentrating on the shore and going as fast as I could, while trying to keep my toes out of the water. Being a surfer, you read about these things happening but I never thought it would happen to me. I have surfed on my own before, but it was really nice having other people around.' Shark expert Clinton Duffy said the encounter with what sounded like a Bronze Whaler shark was unusual: 'I can't recall any similar incident with a Bronze Whaler and surfer in New Zealand – it's more typical of a Bronze Whaler being attracted to a spear fisherman, when they can become persistent and aggressive.' He added it was unlikely a shark the size of the one that had attacked Hobson could eat a person, but it had the potential to inflict a serious injury: 'It sounded determined, but it was more than likely the shark viewed Mr Hobson as a threat, and went into a self-defence mode.'

'THE SHARK IGNORED MY PLEAS AND BUMPED INTO ME...'

In January 2005, Paul Morris was attacked while on a fishing expedition at Taurpiri Bay, Northland. Later, he would write about the incident in detail:

> My first sense of foreboding doom began to creep over me when the mainline moved back to its original position and the nice snapper I was hauling in lay on the surface before me. Most of it was missing – only half of its head was left – and I didn't even feel it happen. It was then, out of the corner of my eye, I saw the large dark shadow moving slowly back towards me. As the shadow turned to shape I couldn't believe what I was seeing – it had to be the biggest shark ever! The giant shark cruised slowly past my kayak. Initially it was almost side on to me and no more

than five meters away. It was swimming in a series of "S" turns. Then, as the shark came closer, I realized that it was two to two and a half meters longer than my 4.15-metre kayak. It also had a fresh deep gash across its tail. I was terrified. I could not take my eyes off it. A tonne or more of hungry shark was now heading straight for me. I grabbed a knife and hurriedly cut the mainline away, hoping to put some distance between me and the still-hooked and struggling snapper on it. In my haste I accidentally slashed a deep cut into my left knee and dropped my knife overboard. The shark wasn't interested in the fish on the line; it was still heading directly towards me. I momentarily closed my eyes as I braced for the first impact. When it didn't come I opened them just in time to see the head of the shark diving under the centre of the kayak. It was now very interested in what was on offer. When it came back, it started nudging the bow of the kayak. It was all too much. Another more powerful wave of fear overcame me, my body was screaming fight or flee now but my mind was saying, stay calm and you might live. When I looked up to take my bearing I often lost sight of the shark, only to have it repeatedly surprise me, usually by bumping into the bow or stern. It seemed as though time had stopped. I was crying my eyes out, certain I was going to die. I even tried swearing at it to stop and leave me alone. The shark ignored my pleas and bumped into me at least 12 times over the next 20 to 30 minutes. I was finally nearing shallower water and the shark somehow pushed or pulled the stem of the kayak under the water. I think it didn't want me to get away. I had the will and begged God to give me the strength to try

> to make it. Then with all my heart and every ounce of power I could muster, I paddled frantically towards the shore. During the night I had cold sweats, later I had sleepless nights plagued with nightmares. I eventually sought counselling and had medication but it was a month before I was getting back to normal.

On 1 February 2010 Lydia Ward, 14, cracked an attacking shark over the head with her bodyboard until it finally let her go when she was caught in waist-deep water at Oreti Beach on New Zealand's South Island. The shark had grabbed her hip but she managed to get away. Lydia recalled: 'I saw my brother's face and turned to the side and saw this large grey thing in the water, so I just hit it on the head with a boogie board – that's what you are meant to do. I showed Dad and he didn't really believe me but then I showed him my wetsuit, with all the blood coming out, and he believed me.'

Following Lydia's tale of amazing self-defence, other similar accounts of close encounters with sharks soon followed. Aaron Muilwyk had a chewed-up speargun to prove his story. Muilwyk was swimming near Cosy Nook when he had a brush with a shark – not once, but about eight times. He was catching moki fish (a sushi delicacy) and blood in the water attracted a shark that he estimated was 2–2.5m (6.5–8ft) long. Muilwyk said: 'I poked it with my speargun and it went away. I thought that was the end of it, but it came back and I lunged at it with my knife, striking it in the snout. It went away again but it kept hanging around. I wondered what is this thing doing?'

The shark came back again and to fend it off, Muilwyk jammed his speargun into its mouth. He was uninjured, but his speargun had become scratched and slashed by the shark's teeth. 'I don't know if it wanted me or my fish but it was pretty full-on. If I had not had my gun, it could have been curtains,' he said.

Aidan Swale recounted his battle with a 2m (6.5ft) shark as he struggled to bring it aboard his boat at Cozy Nook. He had been fishing with a friend for blue cod when the shark appeared. Initially, they thought that the line had got snagged. It took the two men 30 minutes to bring the broad-nosed seven-gill shark to the surface and another 30 minutes to pull it on board a boat with a gaffe. Said Swale: 'Aw, mate, it was big! I thought I'd pulled New Zealand.'

On 24 January 2009, Auckland fisherman Kelvin Travers fled for his life when a shark in a feeding frenzy lunged at his boat and smashed off an outboard motor. Travers had been fishing for tuna near the Alderman islands, about 22km (13 miles) from Tairua and Pauanui. He said: 'I was coming up to a school of fish being worked by birds and I saw this big fin. It went for the fish teasers attached to the stern and it hit the boat with an almighty thump, knocking me across the cockpit. I started madly trying to pull in the lures and the next thing it suddenly launched itself at the back right corner of the boat and I thought it was coming in. I dived for the engine controls and floored the throttle out of there. When I looked round, the auxiliary outboard motor had been ripped off the back.' Travers said that in his haste to get away he did not get a good look at his attacker, but believed it to be a Mako shark in a feeding frenzy among the fish. All that was left of the motor was the handle bearing teeth marks.

Six days earlier, on 19 January, another boat had its outboard motor attacked by a Mako shark. Mrs Bryn Mossman was fishing with her husband David and two friends off Hawkes

Bay. David managed to see off the shark by banging it on the nose with a boathook as it circled their aluminium boat for half an hour or so. Bryn recalled: 'We saw the teeth coming for us following fish we had caught, then it hit the boat and swerved to the side. It circled for about half an hour and was lifting its head, looking at us as it was swirling behind the boat. Then it started to attack the motor. It was more interested in the boat than the blue cod we threw it.'

The body of the boat was undamaged but the Mako left its mark in the outboard motor. It was believed the two boat attacks were because sharks are attracted to outboard motors because of the small electric field they set up in the water, which is similar to shark prey. Shark chaser Boyd McGregor said there was 'easily the highest number of sharks in ten years' in the water that summer – a sign of a healthy marine environment. Mako sharks are common at this time of year and are found around the North and South Islands.

Naturally, the incident was enough to put the fishing party off putting their feet in the water. Said Bryn: 'We often dangled our feet on either side of the kayak to stabilise it while we fish, but after hearing my shark account most would stop doing that – it definitely makes people aware they are out there. I wouldn't be swimming out to sea!'

Cautions for Kayakers

Research has been carried out using a camera attached to a surfboard, which showed the Great White typically stalks its prey by swimming along the bottom and then strikes after launching a lightning-fast vertical attack. This explains one Californian kayaker's report of being rammed, then left alone to swim to shore. He was in the water about five minutes after the attack and when his kayak finally drifted in, no marks were found on his boat to prove the incident. It is possible that the shark might have been more interested if he were fishing and had bait along for the ride or had a bleeding wound.

Along the Pacific Coast, only 5 per cent of the 108 reported unprovoked attacks were on kayakers.

In Hawaii, the shark to watch out for is the Tiger. Oahu has beautiful reef fish to entice the kayak snorkeller – fish that in turn feed on reef fish, a favourite prey of sharks. Sharks are also known to follow kayak fishermen around because of the alluring smell of fish blood.

What kayakers want to know, of course, is just how safe they are in waters where sharks are known to roam. One helpful piece of advice is that those using 'sit-on-top' kayaks rather than 'sit-in' ones are more likely to attract sharks because they may also be diving or fishing, thus causing movement in the waters. Movement attracting sharks is also caused when a 'sit-on-top' kayaker leaps from the kayak in shallow waters before landing. However, a 'sit-on-top' kayaker does have the advantage of jumping back onto his boat if in danger, while the 'sit-in' kayaker would be longer in the water.

The following advice was issued by TopKayaker.Net on how kayakers can avoid shark attacks:

1. Contact your local Fish and Game Department for trends in predator activities, seasonal behaviour

patterns for sharks, what type of sharks to look out for, areas and times of day of increased predator activity or recent attacks. This is usually dusk and dawn as darkness gives any predator an advantage, but don't just rely on that.

2. Good indicators of shark activity in the water are commercial fishermen and seagulls or other sea birds showing frenzied interest in a particular area. The presence of dolphins is an attractive lure, too – they hunt the fish and the sharks join in with the free-for-all. Sandbars, coral reefs and steep drop-offs into the water are also favourite shark areas worldwide. Large schools of fish making evasive manoeuvres like jumping bait fish are likely being chased by one.
3. Shiny jewellery, watches, rings, etc. may appear as fish scales attracting attention. Sharks also reportedly see high contrast in colours often used for safety – like bright yellow – but this is no reason to be unsafe in choosing those items most visible when in need of a rescue. Just avoid looking like a fish. Black wetsuits are typical good choices but for snorkelling or diving off your boat, I might not choose yellow fins!
4. Paddle in groups in waters that appear suspect. In such waters don't let your dog swim or splash overboard as this could appear to be prey activity from below. Avoid swimming off your kayak near river mouths or estuaries with turbulent waters. Avoid swimming with schools of fish as these are often being pursued by sharks.
5. Sharks have an acute sense of smell. If kayak fishing

and filleting your bait as you go, be careful to keep your raw fish scraps or bait secured, not thrown overboard. Kayakers should be especially cautious with bleeding wounds or women kayakers, if menstruating.

Shark expert Dr. George Burgess, director of The Florida Museum of Natural History, also drew up a list of helpful advice for kayakers, if threatened by a shark:

- If you see a shark from your kayak, do not panic. Most likely the shark was attracted by something you were doing or by something in the area of your activity. If you are fishing and have a bait bucket over the side, let it go. If you think some catch of yours is attracting him, let him have it.
- Regardless of the reason for its attraction do what you can to eliminate it and calmly start toward shore, keeping your eye on him, paddling with smooth gliding strokes, not frantic splashing. Gather up close to a paddle buddy as sharks are less likely to go after a group. Stay in your kayak until you reach shore. If you are far from a landing, try to get up against a cliff (in calm water, of course) or wall to minimize the directions he can approach you from.
- Should the shark be making aggressive advances toward the boat, your paddle is the best weapon to discourage him. Hitting him on the snout should work but if he comes back, go for the sensitive gill or eye area. I've wondered why I can't find expert advice on hitting them in the gills or eye to begin with. My conclusion is

you do not want to assault a shark just because it is curious. The snout bump lets him know you are not helpless. Sharks are scavengers, often looking for an easy meal like sleeping fish as a midnight snack, so playing dead doesn't work here. Let him know you have a paddle and know how to use it, but like he and most creatures of nature do, showing ability to do battle is safer than an actual battle for all concerned.

- If he knocks you out of your kayak, hold on to your paddle with all your might. Leap back onto your boat and swiftly, not frantically, paddle into shore. If you lose your paddle or kayak, swiftly, smoothly swim to another kayak or to shore (let the kayak find its own way in, if necessary). If you can't get to shore, find a way to back up against something to again limit the directions he can approach you from. And again, don't play dead. Use your hand to bump his snout, if you have lost your paddle. Leaping onto your kayak swiftly, quickly, even effortlessly from deep water is achievable and an invaluable skill all sit-on-top kayakers should aspire to master.

On 16 January 2009, surfer Tane Tokana was attacked off Otago Beach. He said he saw the shark soon after getting into the water but mistook it for kelp. Half an hour later, it cropped up again. Far from being fazed by the predator, Tokana simply nudged it with his surfboard. 'I actually think it wanted to play – it never told us to go, but we thought it would be a good idea to get out,' he said. Nevertheless, police warned of 'a very large shark' cruising with intent along the beach and asked surfers to stay out of the water. Teenager

Chris Blair needed eight stitches after he was bitten by a Seven-Gill shark at the same beach in 2004.

SHARK ATTACKS AROUND NEW ZEALAND

Otago: Total attacks 9; 4 fatal
Auckland: Total attacks 5; 1 fatal
Wellington: Total attacks 3; 1 fatal
Canterbury: Total attacks 4; 0 fatal
Southland: Total attacks 4; 0 fatal
Hawkes Bay: Total attacks 3; 1 fatal
Taranaki: Total attacks 2; 1 fatal
Chatham Islands: Total attacks 2; 0 fatal
Marlborough: Total attacks 2; 0 fatal
Gisborne: Total attacks 1; 0 fatal
Westland: Total attacks 1; 0 fatal
Pitt Island: Total attacks 1; 0 fatal
Star Keys: Total attacks 1; 0 fatal
Unspecified: Total attacks 6; 0 fatal

Fiji Frenzy

On 17 December 2010, Spanish tourist Jordi Gracia was out surfing in the waters off Kulukulu, Fiji, with 12 others when attacked by a shark. He felt the predator tugging at his ankle and desperately made his way back to shore. Witnesses said the water was turning red around him. Said Gracia: 'My first thought was to get out of the water as soon as possible. I did not feel any pain at first because I was in shock.' He was initially taken to the local hospital and then to Lautoka Hospital to be treated for an open wound to his ankle.

Gracia had been travelling around the world with fellow journalist Albert Martinez and Fiji was their third stop after New Zealand and Australia. Martinez said of the attack on his friend: 'I was really worried about him because when I saw the amount of blood I thought the shark had ruptured an artery. I only focused on getting him out of the water and after we had reached the hospital, I was very relieved. We normally are the ones writing the news, but this time we are the story.' And the incident did not put Gracia off the beautiful island of Fiji: 'This has not changed our views of Fiji. We definitely will return because I believe that if it happened once, it can't happen twice,' he vowed.

Fiji is also home to the Sigatoka River, which in turn is home to sharks and there have been several documented attacks on humans. On 18 March 2006, surfer Paul Chong Sue, 21, was bitten on the arm while paddling out to catch a wave at the mouth of the river. Fellow surfer Ratu Naiqama said: 'The shark came from below his board and bit his right arm. It tried to pull him under but Paul fought back and managed to free his arm from the shark's grip. There were around 15 surfers in the area and they came to his aid...After the attack, the shark kept circling the area. Even though we thought it was going to strike again, we did not pay much attention because we were rushing to get Paul onto the beach.' An Australian nurse, who was also there to surf, helped treat the wounds before Chong Sue was taken to Lautoka Hospital.

CHAPTER 6

SAVAGERY IN THE SOUTH PACIFIC

'SHE HAD ALREADY LOST HER LEG AND THE SIDE OF HER STOMACH...'

It was every parent's nightmare – their little girl taken away from them in the jaws of a shark.

But that is what New Zealand dairy farmers Grant and Sheree Webster had to endure on 22 June 2005, as they watched their daughter Alysha, 7, swimming in shallow water off a beach at Malekula, Vanuatu, in the South Pacific. They could only look on in horror as Alysha went under the water and then re-surfaced. Mr Webster dashed into the water but found his child with extensive injuries: the shark had bitten her left side, taken off her leg and caused severe internal damage. Inspector George Songi of the Lakatoro police said Alysha suffered severe bleeding and would not have survived long after the attack: 'The other kids were in a canoe. The little girl was swimming not too far from her parents; they were sitting on the beach. All of a sudden the shark came in and attacked. The father tried to rescue the daughter but when she came up, she had already lost her leg and the side of her stomach.' He said that Mr and Mrs Webster used a small boat to take Alysha to the mainland, where they flagged down a truck, which drove them to Norsup Hospital, about 7km (4 miles) away. The girl was dead on arrival.

The family had sailed their yacht to Vanuatu from the marina in Whitianga, near to where they live in the Coromandel community of Whenuakite.

Mr and Mrs Webster later flew back home to Port Vila, New Zealand, with their daughter's body on a charter flight organised by the New Zealand High Commission. Paul Willis, New Zealand's High Commissioner in Port Vila, said the family wanted to return home as soon as possible. He said he believed that after the fatal attack, local people were still swimming in the area, adding: 'Sharks are a fact of life in Vanuatu. There are sharks in these waters, but many, many people swim around the islands here so one has to be pretty unlucky, I think. Some areas are known to be more dangerous than others and local custom has it that beaches with black sand seem to attract sharks. The rule of thumb has usually been that if you're in one of the more remote areas and locals are swimming at the beach, they probably know what is safe and what isn't, but in this case that wasn't enough.'

A few days after the attack local teacher Lapen Tillison, 40, said he had twice warned the Webster family not to get into the water – it was advice he always gave to people visiting on yachts. For the first warning he had paddled out in a canoe to Alysha's parents' yacht just after it pulled into the bay. He said: 'They had just dropped the anchor and jumped in for a swim. I went out and introduced myself, and told them it was not safe to swim in the sea. I told them not to swim because here we do not swim in the salt water – we swim in the fresh water, not on the beach.' He added that he later saw the

Webster family get out of the water to have lunch but then they went back in – 'I went back to the boat and heard Alysha ask her mother if she could go for a swim. Her mummy said, "It's not safe" but she went for a swim and after ten minutes, the shark attacked.'

Tillison said local children, including his own, knew it was not safe to go into the water. One good reason for the warning, he said, was that fishing boats came into the bay to wash down and the strong smell attracted sharks. A fishing boat was at the beach doing just that on the day of the attack, he added, before he had invited the family to relax on the shore instead. Tillison said that just two days earlier locals saw a Great White shark in the area: 'It was a very big one – they thought it was the same one.'

Journalist Mark Lowen, publisher of the *Port Vila Presse*, said local people knew where it was safe to swim: 'There are areas where locals know sharks congregate – it's Malekula where most people are taken by sharks. More people are attacked and killed there than anywhere else.' The attacks were often not reported, or reported only in local media, he added.

Department of Conservation shark expert Clint Duffy said Tiger sharks were one of the common species in the South Pacific and inhabited deep water as well as reefs and shallower water: 'They are designed to feed on whatever they come across, from things like rock lobsters to dolphins, other sharks and turtles. They are capable of biting a turtle in half.' But there had been no positive identification of the shark that killed Alysha Webster. Duffy said although fatal shark attacks were rare, tourists should be wary of where they went in the water as sharks were so common; they also hone in on pale skin thinking it is prey. Swimmers were also advised not to swim at night or wear bright clothing or shiny objects, which the sharks could mistake for small fish. Locals avoid some areas at certain times of the year, but one observed: 'In some cases it's just a matter of being in the wrong place at the wrong time.'

At Whenuakite School, where Alysha was a pupil, Principal Jamie Marsden said trauma counsellors had helped Alysha's friends to cope with the horror: 'There have been sad moments, there's lots of sharing and Alysha's room is coping really well. They're having the times when they need to have a cry, to write things down, to put little cards on the desk.'

Three years later, Grant and Sheree, their daughter Jessica, 12, and son James, 8 – who had both been nearby at the time of the horrific incident – were still coming to terms with the attack that had wrenched their lives apart. To honour their daughter's memory, as well as in the hope of uniting their family, they bought a nursery at Hot Water Beach and worked hard to transform it into a holiday destination. Sheree said: 'It's a perfect place for a campground. It has native trees and is metres from the beach. We want people to get back to family; we want to bring families together by having them slow down for two weeks in the sun.'

It was the location for one of the CBS reality television series *Survivor*, but Vanatua in the South Pacific Ocean is listed in the guidebooks for the threat of sharks. Some areas have a reputation for being more dangerous than others, including Port Sandwich, where New Zealander Andrea Rush was attacked in 1992. Despite a punctured artery in her leg, she survived the attack.

'I was swimming on the surface of the water. I didn't

know where I was. It was a moment of absolute horror when the shark actually bit – it was completely unsignalled. I didn't see a fin even; there was no warning. I was being dragged down and this huge thing from underneath had me. As the shark dragged me down, my left hand came down onto it as an instinctive reaction on the nose. I came back to reality, and realised that I was in the water and a shark had bitten me and could come back and get me again. That was the moment of pure terror,' she recalled.

Vanuatu comprises a group of about 80 islands and lies 2,000km (1,242 miles) northwest of Australia.

American aviators fought for their lives after being surrounded by Tiger sharks, having been forced to take to a rubber life raft when their plane went down in the Southwest Pacific Ocean on 12 December 2011. One shark was killed with an automatic rifle. Somehow the men escaped injury, which was incredible considering they were in the water for 34 days before being rescued. Another American aircrew did not fare so well when they crashed in the Southwest Pacific: after taking to a life raft, two of the nine men were killed by sharks.

On 13 May 2000, around 20 or so circling Tiger and Bull sharks prevented the recovery of bodies of three people who died when a Piper aircraft crashed into the sea at Mont Dore in the South Province of New Caledonia – islands boasting French chic, which are located in the

Southwest Pacific Ocean 1,500km (932 miles) east of Australia. Three months earlier, on 15 March 2000, Gilbert Bul Van Minh, 35, was fatally attacked by what was thought to be a Tiger shark while spearfishing at Poum, in the North Province. There were two shark attacks in New Caledonia in 2007; Stephanie Belliard, 23, was killed while swimming in the Bay of Luengoni in the Loyalty Islands on 30 September, and on 25 January 2007, Jesse Jizdny, 30, had his leg badly bitten by a Tiger shark at Kaala-Gomen in the North Province.

CHAPTER 7

AMERICAN SHARK ATROCITIES

'IT WAS JUST RED... THE WHOLE WAVE...'

Surfer Matthew Garcia could only watch in horror when a shark attacked his bodyboarding friend as they enjoyed the waters at Surf Beach, California, on 22 October 2010. Lucas Ransom, 19, bled to death after his left leg was virtually bitten off.

'When the shark hit him, he just said, "Help me, dude!" He knew what was going on. It was really fast. You just saw a red wave and this water is blue – as blue as it could be. And it was just red; the whole wave,' said Garcia, who was torn between getting his friend out of the water and swimming to the shore to find help. The decision was made when he saw Ransom floating in the water – 'I flipped him over on his back and unhooked his arms. I was pressing on his chest and doing rescue breathing in the water. He was just kind of lifeless, just dead weight.'

Officials at the Santa Barbara County Sheriff's Department later confirmed Ransom's death. A Great White was believed to have been the killer, but there was no firm identification.

Surf Beach was closed after the fatal attack. Lucas's mother Candace said she had implored him not to go surfing without studying the area more closely first, but, 'he was so excited. He said, "Mama, there are 10-foot waves!" I was concerned for his safety. He was in an area where he had never been before. He said, "Mom, I'll be fine – I'll give you a call when I get out."'

Touching memorials to Lucas were created on YouTube. A few days after the fatal attack, his friends and family held a traditional memorial service at a local church and in the afternoon gathered on the south side of Oceanside Pier to paddle into the Pacific and scatter flowers. 'I can't explain the feeling out there – it's one of a kind when you're out there riding the waves. It's going to be pretty meaningful for all of us,' said Lucas's brother Travis. His friend Emily Xiao added: 'After everything that's happened I think maybe some people might be afraid to go back in, or some people are terrified about what might happen in the ocean – but the ocean was where Luke loved to be.' In fact, the last-known attack at Surf Beach had taken place back in September 2008 when a surfer reported having his board bitten by a shark.

• On 4 February 2010, Stephen Schafer (38) died after a frenzied attack by a group of sharks when he went kite surfing off a Florida beach. Lifeguard Daniel Lund said he first spotted Schafer from his tower and he could tell the man was in trouble; Schafer was lying on the large sail he was using to pull himself across the water. Lund grabbed his long surfboard and paddled 20 minutes through rough seas, fighting 1.5–2m (5–6.5ft) high waves to reach him. 'I get to him. I'm probably within 20 yards or so from him and there's just a lot of blood in the water,' said Lund, adding that he could see several sharks circling nearby. He pulled the injured Schafer onto his board and began paddling back.

Lund declined to describe Schafer's injuries, but said he was conscious and speaking when they got to the beach and paramedics began treating him. However, in a report later released by the Martin County Sheriff's Office, it was stated that Shafer had bite wounds between 20cm (8in) and 25cm (10in) long on his right thigh and 'numerous teeth marks' to his right and left buttocks. He also had bruising on the inside of his right arm and wounds to his hand that appeared to be defensive wounds suffered while trying to fend off the sharks.

The Sheriff's Office said a lifeguard first saw the man in distress at 4.15pm about 274m (900ft) from shore. Paramedics took Schafer to a hospital, where he later died. Friends said he always followed the 'buddy' system while surfing and were surprised he was in the water alone. 'We always know that sharks are out there – you see them this time of the year,' noted Teague Taylor, a childhood friend who said Schafer had taught him how to surf. 'It's hard to believe that such an experienced waterman would make that one mistake.'

Shark expert Dr. George Burgess said: 'Florida as a geographic entity has more sharks than any other place in the world. At this time of year they are lining up in South Florida getting ready to move north as temperatures begin to warm. The sharks gradually move their way northward and disperse. The message to take home is this is a rare and unusual event. It should put the antennae up for people in terms of, "Yeah, we need to be careful when we enter the sea, but we need to do that every time because we're never guaranteed safety 100 per cent of the time when we enter a wild world."' He added that it should be possible to tell which sharks were responsible from the bite marks. Of the four species known to attack humans in the area, Great Hammerheads, Bulls and Tigers all prefer warm water and in winter will migrate and go into deeper waters. The other species that gather are young Great Whites.

• The remains of Richard Snead, 60, were discovered off Corolla, Currituck County, North Carolina, on 12 September 2009. He had been swimming when attacked by a shark. Medical examination showed that he had been eaten alive.

• On 25 April 2008, David Martin, 66, was killed during an early-morning swim off Solana Beach, California – one of the world's most notorious locations for shark attacks. Super-fit, he was with eight other triathletes at the time of the attack, during which Martin was bitten below both knees. The other swimmers looked back and saw him flailing before he was pulled underwater. Witnesses said he resurfaced, screaming. Several of the swimmers pulled him to shore. According to them, the shark had charged at him and lifted him vertically out of the water with both his legs in its jaws. 'They saw him come out of the water, scream "Shark!" then flail his arms and go back under,' said Rob Hill, a fellow member of the Triathlon Club of San Diego, who was running along the beach at the time.

The attack happened around 7.30am. By 7.49am Martin was declared dead. Members of his family visited the lifeguard station where his body lay. The waters were closed until further notice, but the beaches remained open. Red emergency helicopters continued to scour the area for the shark. 'This is a tragic situation for Solana Beach and the surrounding areas and the county of San Diego,' said Solana

Beach mayor Joe Kellejian, who urged the public to listen to safety officers and stay out of the water. 'The shark is still in the area – we're sure of that.'

A specialist from Scripps Institute of Oceanography was called in to determine what type of shark had attacked Martin because witnesses were unable to identify it. But according to shark expert Dr. Richard Rosenblatt, the attack almost certainly came at the jaws of a Great White: 'The wounds and the attack were most likely the behaviour of a Great White shark. We can estimate by the look and size of the wounds that the animal was an adult approximately12ft [3.65m] to 16ft (4.8m) long. I believe that the shark most likely mistook the group of swimmers for a pod of seals – it's just very bad luck for that man.'

For three days, eight miles of the popular beaches remained closed. Shark attacks are extremely rare in Southern California, but there is a migration of the animals from San Francisco to the Guadalupe Islands, where the females come to give birth.

• On 25 June 2005, Jamie Daigle, 14, was fatally attacked as she swam on a boogie board with a friend, Felicia Venable (also 14), just 180m (200 yards) off a beach in Walton County, Florida. The attacker was identified as a Bull shark. Tim Dicus was surfing near the girls when he heard a scream: 'I was about 200 yards out, just past the second sandbar. When I heard the scream, I turned around and saw one of the girls swimming towards the beach frantically, and the other one had disappeared and there was a big dark spot in the water where she used to be. She was unconscious when I got to the blood pool, so I tried to pull her from the water. The shark had made an attack when I was trying to get her out of the water but it gave me enough time to get her onto the board once he had to come back around to make another attack.'

A statement from the local Sheriff's Office said Jamie was severely bitten on the lower portions of her body. The attack happened at around 11.15am in front of a camping ground near the Sandestin Golf and Beach Resort in the Florida Panhandle. Seeing her friend in trouble, Felicia headed for the shore to get help. Emergency services rushed to the scene but the teenager died as a result of her injuries. Lt. Frank Owens of the Walton County Sheriff's Office said the two girls were unusually far out for boogie boarders and normally only surfers went out that distance: 'Once swimmers pass the first sandbar and drop-off, you will experience more sightings of sharks.' Indeed, a shark had been spotted in area waters closer to shore the day before the attack.

• On 19 August 2004, Randy Fry was killed by a Great White at Westport, California. He was with longstanding diving buddies Cliff Zimmerman and Red Bartley as they anchored their boat, the *Dolphin*, off a large rock. Bartley stayed on the boat while Fry and Zimmerman donned wetsuits, snorkels, weights and masks, and went snorkelling in search of abalone. His friends asked him to look out for sharks or the telltale disturbed behaviour of nearby seals. The two men hit the water. Zimmerman recalled: 'The last thing he said was, "I gotta get a couple of abs. Where are the big ones?" I said: "Right below me." The shark was waiting for him beneath us.'

Despite attempts to escape Fry was instantly killed as the beast bit into his throat. Zimmerman continues: 'It was the most dramatic thing I ever saw in my life – it's just not real. This monster came so fast, it happened so fast and was over so fast you think, "How can this happen?" I yelled "Randy! Randy!"' The shark The shark brushed by Zimmerman, who was just an arm's-length away. Meanwhile, Bartley could not believe the horror he was witnessing from the boat: 'I saw the pool of blood spread across the surface of the water and I knew Randy was gone.' Three helicopters searched the scene, but Fry's body – minus his head – was not found until several days later. Further remains were discovered by a beachcomber on 3 September. It is believed the attacking shark mistook Fry for a seal.

On 19 August 2003, Deborah Franzman, 50, became Australia's first female fatality of a shark attack in living history. She died a victim of a Great White while swimming among seals off the central coast town of Avila Beach, California. The shark, believed to be between 4.5m (15ft) and 5.4m (18ft) long, struck from below, breaking the surface of the water and tearing most of the tissue from Franzman's left thigh. Although no one saw the whole creature, a witness spotted a grey fin in the churning water and the authorities said the nature and severity of the attack left little doubt it was a Great White.

Franzman, who was wearing a wetsuit and fins, was swimming alone but in sight of people on the beach, including about 30 lifeguards. A daily swim before work was part of her normal routine. It wasn't until lifeguard Rich Griguoli and his three colleagues noticed the surf was red with blood that they realised what had happened. Meanwhile, came the shout: 'A shark's got her! A shark's got her!' By the time they reached her, Franzman was floating face down. Griguoli said: 'We all looked at each other and knew we had to get her out as quickly as possible – we did it as fast as we could.'

The victim was declared dead at the scene. She had bled to death after the femoral artery in her groin was severed in the attack. At the inquest, Lt. Martin Basti said her injuries were 'catastrophic and not survivable.' Tragically, it could have been Franzman's wetsuit and fins which sparked the attack, with the shark mistaking her for a seal. A spokesman for the local Sheriff's Department said: 'Her friend on the beach noticed she was swimming with some seals. All of a sudden, the seals dispersed rapidly and a large breach of water occurred in the vicinity. This is very indicative of a shark attack.'

Robert Lea, a marine biologist for the state, said the 'from underneath' attack was consistent with the behaviour of a Great White: 'Shark incidents are extremely rare – sharks have no interest in feeding on humans, but as an ambush predator they may mistake a human in a dark wetsuit for that of a marine mammal. On the surface, she is going to be silhouetted, looking like a marine mammal.'

After the attack, the port authority closed the beach, which is located south of Morro Bay, about 321km (200 miles) northwest of Los Angeles. Shortly after the attack on Ms Franzman, the *San Francisco Chronicle* printed an article on how to avoid similar attacks:

> Along the West Coast, the areas known for the most sharks are Ano Nuevo, Bird Rock near Point Reyes and the Farallon Islands. Don't swim with shark food – specifically sea lions or elephant seals. White sharks do not normally hunt humans but their favourite food is sea lions. Deborah B. Franzman, 50, was fatally attacked by a shark as she was swimming with sea lions just off Avila Beach pier. She was wearing a full-body wetsuit and fins.

If you are wearing a wetsuit and fins and you are swimming with sea lions, you are doing a fine job of imitating shark food.

Ms Franzman was the tenth person to die from a shark attack off the California coast since 1954.

• On 3 September 2001, there was a simultaneous double attack. Russian Sergei Zaloukaey, 28, died shortly after a shark grabbed him at Outer Banks in Avon, off the coast of North Carolina. He suffered massive blood loss. His fiancée, Natalia Slobodskaya, 23, was critically injured, with her left foot bitten off at the ankle. The middle finger of her left hand was also bitten off. She suffered severe injuries to her left side and buttocks.

The couple had been swimming early in the evening in water that was just chest-high and the attack had lasted just two minutes. Mary Doll, spokesman for the National Park Service at Cape Hatteras Seashore, said although no one actually saw the shark, they heard screaming and saw the young couple thrashing about in the water. People rushed from the beach and brought both Slobodskaya and Zaloukaev to shore, where CPR was performed on Zaloukaev. The emergency services were called, but his wounds were too severe and he was pronounced dead on arrival at a medical centre in Avon.

Slobodskaya, who underwent surgery to repair torn blood vessels, was told of his death two days later. She herself was to stay in hospital for a month. Later, sitting in a wheelchair, she spoke of how she was not wearing her glasses at the time of the attack and so she had not been able to see what species had been involved but she said that she had felt the shark's 'sandpapery skin' – a sensation that remained with her long afterwards: 'It was rough, it was disgusting – you know, the skin of a beast.' Slobodskaya said that when the shark bumped her from behind, she thought it was a friend trying to scare her. Then the shark grabbed her and her fiancé screamed: 'It's a shark, swim fast!'

As they swam to shore, the couple tried to fight it off. Slobodskaya recalled the horror: 'It was all around us. I didn't feel much pain at the time because I was in shock. One of my concerns is that Sergei spent some of his energy on saving me – I loved him more than my life. I'm very amazed that I kept my life and he did not.'

Surgeon Jeffrey Riblet said he was surprised that the young woman had survived the attack because of the extent of her wounds. Surgery had included skin grafts and Riblet said her youth and good health had saved her. Experts were divided as to whether the killer shark was a Bull or a Tiger. Doll said sharks constantly swim within the Cape Hatteras shore area: 'It's where they live and feed, but attacks are not normal. It's a safe area to swim with the understanding that when you go into any wilderness-type area, there are risks.' The attack came just two days after another fatality 217km (135 miles) away.

• On 1 September 2001, David Peltier, 10, died while surfing with his father and two brothers off Sandbridge Beach on the southern end of Virginia Beach. They were in water that was only 1.2m (4ft) deep and were only about 46m (150ft) from shore. David's father Richard had spotted the shark and shouted to his three sons. He hauled David onto his surfboard and meanwhile, the other two boys made it back to the beach. The shark brushed Richard's leg before lunging at David, who was freed from its jaws after his father hit the shark on its head. Richard then paddled to shore with his son, where witnesses and lifeguards administered first aid to the boy, but his injuries were too severe and the following morning, David died in hospital. The attack had left a large gash in his left leg and severed an artery.

It was the first fatal shark attack in the United States that year; also the first one reported in the area in about 30 years. The beach was closed after the attack, but reopened two days later. More than 40 EMS (Emergency Medical Services) divers and a jet ski had patrolled the beach; scientists with the city's Virginia Marine Science Museum also flew over it in a police helicopter but did not see any sharks. Museum curator Maylon White said the authorities did not know what kind of shark had attacked the boy although it was likely to have been a Sandbar. These are typically 1.2m (4ft) to 1.8m (6ft) long. According to White, Sandbar sharks are not usually aggressive: 'In many cases like this, the shark is feeding and it's after fish and it mistakes the person for the fish.'

David's family released a statement through the hospital saying they appreciated 'the expressions of concern, sympathy and support they have received from the community and ask that prayers on their behalf continue.'

• On 30 August 2000, swimmer Thadeus Kubinski, 69, was fatally attacked just moments after jumping in the sea at St. Pete Beach, Boca Ciega Bay, off the Gulf of Mexico. The attack happened about 4pm – a time when sharks are actively hunting. Kubinski's wife Anna, who was swimming with him, heard a noise and then saw her husband struggling with the shark. She scrambled out of the water to call for help. When she returned to the scene moments later, her husband was clinging to the dock: he was dead before paramedics arrived.

The couple's son, George said his mother had seen a dorsal fin: 'She couldn't describe it – she just wanted to get out of the water.'

Yoli Pate was floating on a raft in her backyard swimming pool near the beach when she heard the horrifying screams from Mrs Kubinski: 'She was screaming for five to ten minutes. The screams were very eerie, very scary. I ran over there, and her husband was floating, face down, in the water next to the sea wall. Blood was everywhere.'

A tooth recovered from Kubinski's body led experts to believe the attacker was a Bull shark. Dr. George Burgess, director of the International Shark Attack File at the Florida Museum of Natural History in Gainesville, described the chances of anyone being attacked by a shark, much less killed by one, as being 'minuscule', before adding: 'Folks shouldn't be terrified.' In his view, the shark probably mistook his victim for prey: 'My guess is that the shark simply reacted to the splash – humans are not a normal food for them. But it underscores the fact that the sea is basically a wilderness.'

Despite the attack, the beaches remained open. 'This is an isolated incident,' said St. Pete Beach Fire Chief Fred Golliner. 'It's rare – it could be another 20 years before we have something like this again.' Meanwhile, owners of the waterfront homes were determined not to let the unprecedented attack keep them out of the sea. Rob Stambaugh, an island inhabitant for 54 years, said: 'Ever since the movie *Jaws*, we all get a little worked up about shark attacks but every weekend, I'm out on the water and never have I been concerned about the possibility of sharks. You look at the tens of thousands of people who swim around here with no problem – I'm sure I'm not going to let this keep me out of the water.' His wife Debbie said: 'It's a tragedy; a man is dead. But I dive all the time, I see sharks all the time – they're more scared of me than I am of them. I'm not going to stop going in the water.'

As with other parts of the world, America has its shark attack survivors, too:

• On 19 October 2011, surfer Eric Tarantino, 27, was bitten in the neck by a shark at Marine State Beach near Monterey, California. He had spotted the approaching shark – thought to be a Great White – but was unable to escape in time. The creature's strength was described as being 'like a car or truck,' pulling him along. Tarantino was saved by friends, who were able to pull him out of the water and stop the bleeding before he was airlifted to safety. A 48cm (19in) gash was left on his surfboard.

Dana M. Jones, Monterey Sector Superintendent for the California Department of Parks and Recreation, said Tarantino's injuries did not appear to be life-threatening but following the attack, signs would be posted along the area's beaches to advise of the shark danger and recommend beach-goers refrain from water activities for the next week. The damaged surfboard was placed in a police truck at the beach, where it became something of an awesome sight for fellow surfers. Later, upon leaving hospital, Tarantino said: 'I feel really lucky.'

Crunching of Bones in Carolina

The summer of 2011 saw a series of shark attacks along the coast of Carolina. Three people were reported injured, including a 10-year-old girl, who was in shallow water at North Topsail Beach, Cape Hatteras, North Carolina on 4 July 2011. While lying on her bodyboard, Cassidy Cartwright was dragged underwater by a shark (a similar attack on a 13-year-old girl occurred in July 2010). 'It pulled me down and it hurt. I just thought it was somebody messing around and I found out that it wasn't because it pulled me down again,' she said. Cassidy's mother Carolyn and a friend rushed to help her daughter: 'Together we got her out, but when we pulled her out of the water, her leg was wide open and it was just a lot of blood.'

The last fatal shark attack in Cape Hatteras was in 2004.

On 19 July 2011, 6-year-old Lucy Magnum was attacked by what was believed to be a Tiger shark at Ocrakoke Beach on the Outer Banks. She was twice bitten on the legs and suffered serious damage to calf muscles and Achilles tendon. 'I heard her scream and I immediately turned towards her and at that point saw the shark right next to her. I ran over to her and at this point it hadn't really crossed my mind that she had been bitten – I just wanted to get her out of the water. We knew right away that this was pretty severe. It was pretty surreal,' said her mother Jordan.

Lucy's father Craig was further out in the water when the attack happened. He said: 'I ran over and we got her up on the beach, and we had Jordan move her hand and the entire lower leg filleted itself open.' The child was flown to hospital and luckily, despite many stitches her leg could be saved. Astonishingly, the brave little girl decided that although she had said she hated sharks after the attack – 'I like dolphins better' – she had now forgiven her attacker. Her mother was equally forgiving: 'We talked about it and decided it was a mistake – the shark didn't want to eat her and that's why he just bit and left it there and swam away.'

On 11 August 2011, 45-year-old Don White was bitten by a Bull shark while swimming off a boat about 4km (20 miles) from Beaufort Inlet in North Carolina. He sustained severe injuries to his right leg, which needed 120 stitches. Witnesses used T-shirts and towels as bandages until help arrived. White said of the attack: 'All of a sudden I felt severe pressure like a strike on my right foot.' His son Donnie pulled him back into the boat. The two said they had seen several sharks circling at the time. White was left with a shark tooth embedded in his leg, which was later removed and kept as a souvenir.

Biologist Andy Dehart says the murky water around the North Carolina shore is often to blame for unprovoked shark attacks in this area: 'The shark sees a flash of pale skin, which has a high contrast in the dark, murky waters and oftentimes that can confuse sharks a little bit. They bite down thinking they are biting a fish but it's a person.'

Courting Shark Controversy

In September 2011, the animal welfare group PETA (People for the Ethical Treatment of Animals) launched a controversial campaign with adverts showing a shark biting a man to death accompanied by the slogan: 'Payback is Hell. Go Vegan'. The group said it had been prompted to launch the campaign after the latest attack on a man called Charles Wickersham on 24 September. The 21-year-old survived the Bull shark attack, which happened while he was spearfishing with friends off Florida's Anna Maria Island. Wickersham's upper thigh bone was exposed and he needed 800 stitches and two operations after being airlifted to hospital. He recalled: 'At first I thought a friend had tugged on my leg. Then I was sitting there, catching my breath and all of a sudden it hit me. The teeth were so sharp I didn't even feel it.'

PETA said it had been compelled to launch its campaign after the incident because 'Humans hook, spear, maim and kill fish for "sport" every day. The most dangerous predator of all is the one holding the fishing rod or standing at the "all you can eat" seafood buffet.' Ashley Byrne, PETA campaign manager, commented: 'With the recent shark attacks in the news, we thought it was a good time to bring this discussion up that will hopefully save lives, both humans and animals. The intent of this campaign is to make the point that the deadliest killers aren't sharks, but humans. Sharks are not the most dangerous predators on earth – we are. Americans alone kill billions of animals for food every year, including fish. And while sharks are natural carnivores, people can choose what they eat.' But she added: 'We're very glad that Mr Wickersham is going to be OK, we just hope that after this painful and frightening experience he'll consider what fish feel when they are impaled and suffocated to death.'

Wickersham's mother Ella retorted: 'I'm not even going to dignify them with a response! It's not even worth my response. They are over-the-top – if they don't want to eat meat and fish, good for them. You can do whatever you want, and I'll do what I want!'

- On 30 May 2011, Kori Robertson, 22, received a ferocious bite to her leg as she waded in water near the beach at Follett's Island, Texas. 'All of a sudden, I felt something jerk me and bite,' she said. 'I was kind of like swimming and thinking, "I just don't want to get bitten again." I really didn't want to look at it – I've never seen anything like that before and it hurt really bad.'
- Kimberly Presser, 37, had to have 150 stitches to an arm wound after a shark attacked her as she waded in water off Mickler's Landing, St. Johns County, Jacksonville on 20 August 2010. She had raised her arm to fend off the beast. Shortly before the attack, she had told her 6-year-old son: 'Sharks are out deep and there's nothing to be scared of.'
- Surfer Clayton Shultz (20) needed 300 stitches to a foot wound after being attacked off Jacksonville Beach, Duval County, Florida on 23 July 2010. 'All I really felt was teeth and tearing. I was hopping back on my board, and the shark came up and grabbed my foot and shook it around a little it and let go. I popped right up, and got on my board and lifted my foot out of the water – it was torn up real good,' he said.
- On 16 July 2010, Charlie Gauzer received serious injuries

to his foot while fishing off Galveston Island, Texas. The bite narrowly missed an artery, but severed his Achilles tendon. 'I started to feel pain. I looked down and I had blood coming out of my leg – it was gashed pretty good, down to the bone,' he recalled.

• Mike Seymore, 49, was badly bitten on the calf as he tried to release a shark at Tarpon Springs, Florida, on 30 May 2010. Involved in shark research, he was attempting to remove a hook from a Lemon shark. He said: 'I took the hook out and he jumped the wrong way, or I moved the wrong way and he got a hold of my leg. That's it. I got careless, he got me.'

• On 6 December 2009, Bob Large, 59, was badly mauled on the calf when he accidentally bumped into a Sand Tiger shark while diving in the Adventure Aquarium in Camden, New Jersey. 'This was an accident – the uncertainty of animal behaviour is always a risk,' said Greg Charbeneau, the aquarium's executive.

• Dan Callahan needed 130 stitches to a foot wound after being attacked by a Bull shark while swimming off the Florida Keys on 26 September 2009. 'Everything probably took less than 30 seconds,' he said. 'I dived in and I got hit from behind on my right foot – it really felt more like a hammer or a sledgehammer. It starts to pull me down. I turn and see the shark with my foot in its mouth and with my left leg, kick it with all of my might. I know a little bit about sharks, so I knew I needed to get loose. I hit the shark on the nose. It released me for a couple seconds. I took two or three massive strokes over to the seawall and, without even thinking, climbed up a barnacle-encased ladder. That immediately puts a nice big barnacle through my left foot, but I didn't even feel that.'

• On 12 September 2008, Daryl Zbar sustained devastating injuries after he was attacked while surfing at Hutchinson Island, St. Lucie County, Florida. His thumb was virtually torn off. 'I got to the outside and as soon as I plopped down on my stomach and dug my right hand in to paddle, it hit me,' said Zbar. 'The thing I remember most is this like, vibrating noise and I think that was his tail flipping back and forth as he was shaking his head. It grabbed me right on the hand and shredded it. I knew right away it was shark on my hand and when I pulled my hand up, he let go and swam away.'

• Bettina Pereira had a very lucky escape on 21 June 2008 after being knocked off her kayak by a Great White at Catalina Island, Orange County, California. The impact sent her screaming into the water. At one point she actually stepped onto the shark. Describing the attack, Bettina's husband Andrew said: 'It came right under her kayak, threw her in the air and threw the kayak in the air. When she landed, she landed on the back of the shark on her two feet – it was incredible. It seemed like a long time but it all happened so quick.'

Miraculously, she escaped with no injuries. Shark expert Ralph Collier commented: 'These sharks have been at the island for millions of years so the fact that something has finally happened over there does not surprise me – I'm surprised it's taken this long.' He described the incident as an 'investigation attack,' adding: 'The shark was not interested at all in eating her, otherwise it would have stayed in the area and done so.'

• On13 June 2008, John Vasbinder (59) was surfing off Cocoa Beach, Florida, when a shark attack left him with the skin on his hand hanging off. 'I never saw it coming,' he admitted. 'In 40 years of surfing, this is my first scratch. A couple of my surfing buddies came over and I say lightning doesn't strike twice, so I'm good for another 40 years.'

• Nine-year-old William Early was bitten on the biceps and lower arm while surfing off Hammocks Beach State Park, Bear Island, North Carolina on 26 May 2008. He had been on a school field trip. The boy underwent surgery at Wilmington Hospital and was allowed home soon afterwards.

• On 28 April 2008, David Alger (18) lost a chunk of his foot after he suffered a shark bite while surfing off New Smyrna Beach, Florida. 'It just latched on to me and two chomps real quick. I kind of kicked my foot away and it just swam away,' he explained.

• Carolyn Griffen, 58, sustained deep wounds to the leg from a Bull shark as she waded off New Smyrna Beach, Volusia County, Florida on 8 September 2007. 'It wasn't like a really bad thing – I was really fortunate,' she said.

• On 27 August 2007, surfer Todd Endris (24) and a friend of Eric Tarantino (above) was attacked by a Great White in the same location. He was with another friend – Brian Simpson – and the two had stopped to watch a playful pod of dolphins. Endris sustained injuries to his back and shoulder and said that when he was in hospital being treated, 'You could see my lungs when they were sewing me up.' In fact, other organs and his spine were exposed by the serrated tooth bite of a Great White.

Todd said he had been paddling parallel to the shore when he was violently attacked from below. Later, he went on to describe the attack in great detail:

> I had just caught a wave and paddled back out. There were four other dudes surfing with me and I was looking

back at them. I paddled right to the outside of them on top of the sandbar and I was sitting parallel, facing south and it hit me coming from the shore, from where they were; there was no pain on impact. There was a bottom jaw underneath my board and the top jaw pretty much like, clamped on my thigh. I instantly started hitting the thing with the butt end of my left fist because I couldn't hit it with anything else. I was hitting it on the side of its head above the eye. It was so powerful and graceful, and it was so fast and effective.

He lifted me out of the water and bit down twice on me – once while I was in the air and once while I was going back into the water. He bit the same area, like an inch away, and gave me another row of teeth marks. For about half a second I was out of the water probably. It was fast; it was swift and powerful. Two bites and he let me go and then I was thrashing around in the water afterwards, and he was underneath me and I was trying to get away from him and I kicked him real hard with my left foot. It was gnarly and then I caught a wave in under my own power. There was a dude out there named Joe, who was yelling at me after the attack to get my board. That seems obvious but I was pretty disoriented. He was paddling away from me and I can't blame him but he yelled at me to get my board, and I did and that was how I got to shore. That's one thing I wanted to clarify.

My board was pretty thrashed – it looked like someone had taken a cheese slice along the bottom of it. My buddy Nick works for the abalone farm and a couple hours before I got hit, he was about three miles directly out from Marina (they collect drifting kelp to feed the abalone).

They were going like, 20mph and his partner said, "What the hell is that?" There was like, a 14- or 15ft White shark just sitting there, basking at the surface. They pulled up next to it and measured it out at about 14ft, so when I got attacked, I got to shore and my buddy Brad called Nick and said: "Todd got hit." And Nick was like, "Holy shit! I saw the shark that hit him three miles out, just basking."'

Todd also had another part of his amazing story to tell: how dolphins may have tried to come to his aid. 'You know, the funny thing is that there were dolphins really close to us the whole time we were out there. They were swimming around us and swimming in front of, and underneath us in a couple feet of water. And after I got hit – apparently from what the other guys said – the dolphins all swarmed around behind me as if they were protecting me.'

Witnesses also described the amazing dolphin behaviour: 'Six dolphins began a defence. They leaped out of the water around Endris, creating whitecaps. They surrounded the victim, slapping their flukes as the water turned red with his blood. Then they leaped right over him, creating a spectacle – one which obviously had the desired effect on the Great White as it soon released Todd's leg. The dolphins then positioned themselves between the shark and the surfer.'

Astoundingly, Todd survived to tell the tale despite the shark having missed his aorta by only 2mm (1/16in), losing half the volume of blood in his body and requiring over 500 stitches and 200 staples. Todd had been swimming in one of the areas that constitute the Red Triangle, known for its abundance of sharks. There are also many sea lions and seals – a favourite meal for the Great White Shark.

• On 27 August 2007, a 24-year-old unidentified surfer punched at a shark at Marina State Beach, Monterey Bay, California. He was heard to scream before the beast pulled him under the water. California State Parks spokesman Loren Rex said: 'Witnesses saw a lot of thrashing and some blood coming up. Other witnesses saw the shark let him up before biting him one more time.' One witness said the shark was at least 6m (20ft) long, but rescuers were not able to immediately confirm, commented Rex. Surfers pulled the victim to shore and administered first aid, using a surf leash and a blanket as tourniquets to stop the bleeding until rescuers could arrive.

The victim was airlifted to hospital with bites to his torso and thigh, but recovered after surgery. Marine biologist Steven Webster said the shark probably thought the surfer was a seal: 'They don't come in close just to get a hold of people – this is more often than not a case of mistaken identity.'

• Ashley Silverman, 19, was attacked by a Bull shark on 7 August 2007 after jumping into the water off Islamorada, Florida: 'I jumped in the water and then I just felt it snap on my arm. It was quick, really quick. Then I just felt my hand and felt my bone – I saw my bone. There were pieces of my flesh in the water. Muscles were like, oozing out. It was really gross.' Unsurprisingly, her wounds needed reconstructive surgery.

• In September 2006 there were several shark attacks off the notorious New Smyrna Beach: Christopher Duncan, 33, on 3 September, Dennis Macy, 52, on 11 September and Chris Andres, 19, on 29 September. All were left with serious injuries.

• Dave Marcel decided to kiss a Nurse shark while diving off Key Largo, Florida, on 15 August 2006 and lost his lips. Marcel's somewhat complacent attitude is surprising: 'You kind of pick them up, rub their belly, scratch them, hug them

– you might as well give them a smooch while you're out there.' He had done so many times before but on this occasion the shark decided to kiss back. Miami cosmetic surgeon Dr. Mike Kelly said the injury was a first for him: 'Dave's lip looked like it had been put through a meat grinder or a garbage disposal. I mean, it was a hundred little pieces! Basically it was a matter of completing the jigsaw puzzle and putting the pieces back together again.'

Marcel seemed to have learned his lesson, saying: 'Don't kiss a Nurse shark while it's upside down.'

• On 31 July 2006, a group of people were bitten while petting sharks at Newport Aquarium, Virginia. The exhibit has almost 20 sharks, including young and adults, for visitors to touch. All were said to have been selected for their, er, friendly behaviour. Aquarium biologist Scott Brehob defended the practice, saying, 'What's the reaction to people touching them? They do not mind at all. I've had kids that are anywhere from 8 months to 2 years to 3 years touching these animals – the bites are more like the lines of a catfish. They are bristle-type teeth – it's like a broom.'

• Bruce Lurie had his spinal cord snapped in two when he was attacked by what was thought to be a Tiger shark while 'boogieboarding' off the Southern Californian coast on 29 July 2006. He was left paralysed. Experts believe the shark had mistaken Lurie for a seal as a colony had been seen in the area. He wasn't actually bitten, but the shark's force did the damage. 'I suffered an immediate and catastrophic spinal cord injury,' said Lurie. 'I would've drowned, but my son pulled me out of the water and a passer-by on the beach who knew CPR managed to resuscitate me.' Incredibly, three years later, Lurie was back on his feet.

• On 27 June 2005, Craig Hutto, 16, lost a leg at Cape San Blas, Florida. The teenager was fishing in waist-deep water about 18m (60ft) from shore with his brother Brian and a friend when the shark grabbed him in the right thigh, nearly severing the leg. It was Brian who desperately fought and punched the creature off. Witness Karen Eaker said: 'Within five seconds it was obvious something was wrong. We had heard the word "shark" and then we saw the red water and the tug-of-war going on between him and the shark.' Bill Pascoe was collecting shells with his five-year-old son when they heard the terrifying shouts: 'One man jumped in and kind of looked like he had it in a head lock and was punching it on the head to get him to let loose.'

A doctor, who happened to be nearby, began treatment when the boy was on shore and managed to stem the flow of blood. Hutto was then taken to Panama City's Bay Medical Center, where his leg was amputated. Dr. Reed Finney, a cardiovascular surgeon at the hospital, said he had suffered irreparable damage to blood vessels and nerves between hip and knee, as well as to most of the surrounding muscle. Heartbreakingly, when Hutto woke up in hospital he begged his mother not to let the doctors take off his leg, even though he knew this was already the case.

Incredibly, Hutto literally got back on his feet with a robotic leg. He was a guinea pig for one being developed at Vanderbilt University in Nashville: it was the first prosthetic with powered knee and ankle joints that operate in unison. An amazingly brave Craig said: 'A passive leg is always a step behind me – the Vanderbilt leg is only a split-second behind.'

• On 4 August 2004, James Tiffee, 47, was paddling back to shore at St Petersburgh when he felt himself being lifted out of the water by a shark. It bit into his lower back and grazed most of the left side of his body. Said Tiffee: 'I felt the impact on my back and was lifted ever so slightly into the air. I realised pretty quickly it had to be a shark. I didn't see it, I just saw bite prints and I was immediately concerned with getting in to shore.'

Afterwards, he underwent a four-hour operation for wounds to his back and also suffered bites to his face, buttocks and arm. However, he was very accepting of the incident, saying: 'I was swimming at feeding time and was splashing my legs around – I couldn't have done anything worse but to have slit my wrists in the water.'

• Aaron Perez, 11, put into action what he had gleaned from watching wildlife programmes when he was attacked by a Bull shark at a beach near Freeport, off the Gulf of Mexico, on 25 July 2004. He knew you had to punch at a shark's gills and doing just that saved his life, despite undergoing a four-hour operation to re-attach his arm. Aaron and his father Blas also thrashed the shark with their fishing rods. Aaron said: 'I was watching TV the day before and I saw what to do on the Discovery Channel.' It was thought the shark had been attracted attracted to a bag of caught fish that Aaron had tied around his waist.

• On 7 May 2003, Gerald Gaskins, 34, jumped off his surfboard at the Ponce de Leon Inlet, Florida only to land on a shark. 'I never saw it, but I could feel it spin around and lag me,' he said. 'It came down on my left foot three or four times in about half a second.'

Dr. George Burgess, director of the International Shark Attack File at the University of Florida, said the incident did not surprise him: 'This is the time of year sharks are going to be seen with regularity along the beach. When you surf near an inlet, you are taking a risk.' The previous month, surfer Stephen Flowers, 18, was bitten on the left ankle in the same location. He had to undergo two operations and said: 'It seems to happen all the time in that spot – I knew the sharks were there when I went into the water but I figured it wouldn't happen to me.'

• In August 2001, no less than seven surfers were attacked off the sharks' paradise of Smyrna Beach, Florida.

• On 6 July 2000, Ashley Walker, 12, was playing in the waters off Pine Island, Corolla, North Carolina, when she was attacked by a shark up to 5m (16ft) long, believed to be a Tiger shark. She suffered lacerations to her calf.

In February 2008, fishermen from the Royal Seafoods company at Monterey, California hauled in a 4.8m (16ft) shark – the biggest they had ever seen. But that was not their only surprise. As they hoisted the shark (described as a Cow shark) up, a wallet fell out. John Pennisi, the ship's captain, said: 'I have seen a lot of sharks and a lot of things but never a wallet coming out of one.' The wallet was inscribed with the date 1972 but had no identification; bits of paper had writing in English and Japanese. The shark's body also contained several seagulls and 'a few furrier sea creatures.'

CHAPTER 8

BEASTLY SHARKS OF THE BAHAMAS

'WE WERE GOING TO CUT THE HOOK OUT OF HIS MOUTH AND LET HIM GO WHEN HE REGURGITATED A HUMAN FOOT...'

Austrian lawyer Markus Groh, 50, was justified in feeling apprehension about the shark-bait trip he had been invited on: he was fatally wounded on 24 February 2008. While diving alone on a remote coral reef, Groh was savaged. He had rightly been nervous about the six-night trip with nine other Austrians on the charter boat *Shear Water* to scuba-dive with sharks in the Bahamas for the waters would be strewn with dead fish to attract a host of sharks, including Tiger sharks, Lemon sharks, Hammerheads and Bull sharks. Furthermore, the divers would not have the protection of cages as they watched the beasts circle; some said this would be the experience of a lifetime, others that it was simply dangerous in the extreme.

The whole 'adventure' was indeed fraught with deadly potential. Groh and two others dived to about 24m (80ft) below the surface and swam around a 'chum box' (food crate) containing pieces of fish. It was not long before the Bull sharks, rightly deserving of their fearsome reputation, came calling and they did not take kindly to what has been described as humans not only 'trespassing in their underwater kingdom' but also calmly taking photographs as the predators wove in between them.

Dive master Grey O'Hara later joined the three divers with a new supply of bait. Suddenly a 2.1m- (7ft) long Bull shark bumped the chum box with its snout, nudging it to within less than a metre of Groh. Although anticipating the shark's next action, O'Hara was not quick enough. Within moments, a cloud of sand was swirling up as the shark sunk its teeth into Groh's left calf muscle, slashing through arteries and veins. In an effort to shake off the shark, Groh rolled onto his back. O'Hara desperately grabbed Groh's oxygen tank and kicked the shark several times. The shark released Groh and swam away, leaving its victim in a bloody pool.

O'Hara rushed the injured man to the surface, where he passed out on board the boat. The crew and fellow divers swaddled him in blankets and raised his mauled leg above his heart to try and stop the bleeding; they also poured a coagulant powder into the gaping 20cm (8in) wound to staunch the bleeding. At one point, Groh's heart stopped and CPR was carried out. The boat's captain, Jim Abernethy, radioed the US Coast Guard for help. It would be some time in coming, for the *Shear Water* was about 80km (50 miles) off Fort Lauderdale, Florida. The wait for help was to last nearly an hour. Groh was then airlifted to Ryder Trauma Center at Jackson Memorial Hospital, where he was pronounced dead on arrival. The County Medical Examiner said death was caused by 'exsanguination' – bleeding to death. Groh had 'passed away from his injuries sustained by a shark bite,' said Coast Guard spokeswoman Petty Officer Jennifer Johnson.

In the diving world, the *Shear Water* is well known for its shark diving trips, including baited dives in remote areas of the northern Bahamas; Groh was savaged by the Bull shark at a site called 'The End of the Map'. The horrific incident sparked sick jokes on the internet and understandably, also renewed calls to ban shark diving in the Bahamas. Dr. George Burgess, director of the International Shark Attack File, said that although this was the first fatal attack involving shark feeding, there had been two dozen or so attacks on divers taking part in the practice. Burgess said: 'I oppose shark feeding, not so much because it's dangerous but because it trains sharks to expect food from people and not to fear them. They lose their natural caution around human beings. For the same reason on land, you don't feed alligators, you don't feed raccoons and you don't feed bears. It's changing the behaviour of sharks and changing the ecology of these areas by concentrating sharks in one area.'

It was later revealed that boat owner Jim Abernethy had been told by the Bahamas Diving Association to exercise caution with aggressive shark species. Comprising 36 charter businesses, the Association had sent a letter in the last year to Abernethy and other dive operators to stop 'open-water non-cage Shark Diving experiences with known species of potentially dangerous sharks. Some dive operators have chosen to disregard standard safe-diving practices as it relates to interactions with Tiger sharks and other potentially dangerous species of sharks, in various locations within the waters of The Islands of The Bahamas.' Association president Neal Watson said: 'Most operators do a safe dive behind cages, but Abernethy, for whatever reason, simply refused to comply with the safe diving practices in violation of our standards in the Bahamas. It is tempting to offer a cageless option to customers; both the photographs and personal experience would be enhanced. But the risks are too great. Him working with Tiger sharks and Bull sharks uncaged is totally irresponsible and dangerous. There's not a shark expert in the world that would put divers in the water, with chum specifically to attract Bull, Tiger and Hammerhead sharks without a cage – that's putting people's lives at risk. It wasn't a matter of "if", it was a matter of "when."'

The attack prompted an online posting from one dive group: 'We normally would not cover news about a shark attack. We leave the sensationalizing to the mainstream media, who have done a great job of giving these beautiful creates a really bad name. However, yesterday's incident touches the hearts and souls of the underwater imagery community. The online forums are ablaze with conversations regarding the fatal shark attack on the *Shear Water*, a popular shark diving boat among underwater photographers and videographers, based in Ft. Lauderdale and operating in the Bahamas. Our condolences go out to the family of Markus Groh, who was attacked by a Bull shark and subsequently passed away.'

A 'Blog' written by Abernethy just a week before the fatal attack on Markus Groh reported: 'Last week we figured out a very unique way to capture much better images of the wide-open mouth of the Tiger shark and this week's images are quite impressive, showing the success. We will be adding this to all of our future trips. It is quite a pleasure for me to constantly improve the way we do things here at JASA (Jim Abernethy's Scuba Adventures) from a photographic perspective that allow our repeating guests to constantly come back with better and completely different shots. I have quite simply been blessed with the most wonderful crew that truly loves their work and are constantly making everything better for everyone here. Without all of them, I would quite honestly not have the success that we do have.'

On 26 January 2010, Jim Abernethy was attacked by a shark and bitten in the arm. He was on a week-long shark expedition on the *Shear Water*, diving for Tiger and Hammerhead sharks – a trip scheduled to return on 27 January. Abernethy was reported to be conscious and responsive when he arrived at hospital for treatment. Later that day he issued a statement through a friend:

> Thank you for your concerns and well-wishes tonight. I'm going to be fine and so is the shark. My concern is for the future of sharks. Each day more than 200,000 are killed – mostly for shark fin soup.
>
> Since the 1970s, shark populations have plummeted by as much as 80 per cent, with some species reduced by a staggering 97 per cent.
>
> I've spent the last two decades of my life in the Bahamas with the sharks that I love.
>
> Today's minor incident will not deter me. I will continue my mission to help protect these beautiful animals and share their grace and wonder with others.
>
> I am dedicated to promotion and establishment of marine preserves for these creatures and all marine life. This is the only way to help protect the future of our oceans – and the life, resources, and joy it brings to all of us.
>
> Thank you again for your concerns and well wishes.
>
> I plan to be back out to sea in a few days.

- On 6 September 2010, the remains of Judson Newton, who disappeared after a boat trip off Jaws Beach, in the Southwest Providence, were found in the stomach of a 3.6m (12ft) Tiger shark. Newton and his friend, Franklin Roosevelt Brown, had been reported missing after they failed to return from the trip on 29 August. Three other men who had been with them were rescued. The discovery of human remains in the shark, caught off Nassau, was a shock for deep-sea fisherman Humphrey Simmons and his two companions. As they went to cut the hook from the unusually heavy shark's mouth and let it go, it regurgitated a left leg. Said Simmons: 'Everything was intact from the knee down. It was mangled, but there was still flesh on the bone.' He had called for one of his fellow fishermen when he realised how 'extra-heavy' the shark was; the men shot it. 'We tied the rope around his tail fin and pulled him towards the boat. We were going to cut the hook out of his mouth and let him go when he regurgitated a human foot, intact from the knee down,' he continued.

  The fishermen decided to take the left leg and the shark to Nassau. 'There was so much stink coming from the shark's belly, and the belly was so huge that we thought that there might be more bodies inside,' explained Simmons. Fortunately, they came across a Defence Force boat and flagged it down. The shark's body was offloaded, cut open and inside were the remains of a headless man – a severed right leg, two severed arms and a torso in two pieces. Assistant Commissioner Hulan Hanna confirmed the discovery but said it was not known if Newton was alive or dead when he was eaten.

- On 24 May 2008, two identified men who died after their boat capsized in a squall off West End, Grand Bahama Island, became the victims of what were believed to be Tiger sharks. Their bodies were seized upon and eaten.

As already shown in this book, many survive encounters with sharks and the Bahamas is no exception when it comes to lucky escapes:

- On 2 October 2010, Jane Engle was attacked by what was believed to have been a Lemon shark while surfing with her husband Ronald and some friends in the area of Elbow Cay, Abaco. She sustained bite marks between her left ankle and knee, and required some 75–100 stitches. Ronald Engle described the shark as yellowish brown and between 1.5m (5ft) and 1.8m (6ft) long. He said of the attack on his wife: 'Luckily, we had a couple of my buddies there who helped transport her – we had to put her in a boat and then get her to a car, then on to the medical centre. Luckily, everything came together; there is no threat of her losing her leg or anything. Luckily, the shark bit down a couple of times and let go. The wound is bad, but it could have been a lot worse. There has never, in recent memory, been any shark attack here. Sometimes these smaller sharks are a little more aggressive but we have surfed this whole area for the last 30 years and have never had a problem before – we think it's an isolated event.'

Engle also criticised the Bahamas' shark-baiting practices, adding: 'Personally, I would discourage any shark-diving adventures in the Abacos or the Bahamas, where they literally feed sharks to bring them in. I don't think that's a good thing for the Bahamas because these sharks can interact with other humans someplace else and expect to see food.'

- On 6 May 2009, Luis Hernandez, 48, was attacked and seriously injured by a Bull shark as he went diving and spearfishing in the Exuma Islands. It was only the fast actions of his wife, Marlene (46), who lifted the anchor and pulled her husband out of the water that saved him from death. She then made a tourniquet to stem blood loss and got the boat back to land. Hernandez had just speared a grouper at a local reef when he spotted the shark. He said: 'The first thing I thought was, "Wow, nice shark!" So I swam a little closer and thought about spearing it, but decided to let it go. I just poked it so it would get out of my way.' But the shark wouldn't leave. Eventually it lunged at him, sinking its teeth into his arm.

  'I saw them in the water and I knew immediately it was a shark. It was like a nightmare, like a movie,' said Marlene. The shark eventually let go, taking with it a chunk of Hernandez's forearm; he could see strips of muscle dangling from his bone. Later he had to endure several operations on his arm, but was full of praise for his wife, saying:

'This is a woman who can barely stand the sight of blood but she came through that day – she became my angel!'

- On 11 January 2006, Hayward Thomas and Shalton Barr were intimidated by a 3m (10ft) pregnant shark while fishing for lobster off Sandy Cay, Grand Bahama Island. Taking no risks, they killed their potential attacker.
- On 8 April 2001, Wall Street banker Krishna Thompson, 36, was left in a critical condition after being attacked by a shark at Freeport, Grand Bahama. His wife, Ave Maria Thompson, watched the attack. She said: 'He was just swimming off the beach when something, a shark, grabbed his leg and started pulling him down. He kept punching and punching. He has cuts on his hands because of that.' Thompson had his leg amputated just above the knee and suffered severe blood loss.

  The couple had been celebrating their 10th wedding anniversary. Incredibly, Krishna Thompson went on to join the shark 'saviour' group, supporting a ban on sharks being hunted.

SHARK SPECIES OF THE BAHAMAS

- Blacktip
- Spinner
- Sandbar or Brown shark
- Blacknose
- Nurse shark
- Lemon shark

CHAPTER 9

HAWAIIAN SHARK HAVOC

'IT LOOKED LIKE A BIG, STUBBY SUBMARINE...'

On 30 August 2008, Kameron Brown, 27, was killed by a Great White shark at McKenzie State Park, where he had been drinking before jumping into the ocean just before dark when he got into difficulties. He tried to climb onto some rocks in rough surf and fishermen on shore attempted to throw him ropes but Brown never made it back. Bystanders last saw him drifting south out of range of the flashlights that the search parties were using on shore. The search was called off after rescuers found some of his clothing and a battered bodyboard that appeared to have been hit by a shark. A pair of white shorts had teeth marks in them, reported Battalion Chief Darren Rosario, providing 'strong evidence' that a shark had hit on Brown – 'They had teeth patterns like a shark bite,' he said, before adding that a large white shark had been seen in the area earlier in the day. It was described as 'very big'.

Chadwick Chun Fat, one of the rescue team, said: 'The shark was about the size of a 25ft [7.6m] boat. We thought maybe it was a big Tiger shark at first, but then we flew at it with the sun at our back and said, "That's no Tiger shark." It had no stripes, just dark grey; it looked like a big, stubby submarine. We see a lot of sharks out here – Galapagos, Tigers – but that's the first time I've seen a Great White in person.'

On the evening of the attack, around ten fire fighters searched by boat and on shore for Brown's body. A fire helicopter joined the search the next morning and five divers took three dives, but no trace of Brown could be found.

- On 23 February 2006, diver Anthony Moore, 45, sparked a big search when he went missing at Makena, Maui. Diving gear and partial human remains indicating a shark attack were later recovered from the ocean. Maui Fire Department acting Battalion Chief Jack Williams said that a kayaker spotted the diving gear and a small portion of remains on the surface at a popular diving spot known as 'Five Graves' due to burials onshore. Fire department divers went into the water about 457m (500 yards) from shore and found more remains on the bottom, he added.

  Moore's wife was later tasked with identifying his personal possessions. However, an autopsy did not conclusively result in the cause of death being dying from a shark attack. Although the body had obviously been eaten by a shark it could have happened after his death. Those in the area at the time said they had not seen any sharks circulating, but as a precaution the state Department of Land and Natural Resources closed a two-mile stretch of coastline from the Kea Lani resort to Pu'u Ola'i.

- On 7 April 2004, surfer Willis McGinnis, 57, died from shark injuries to his right thigh and calf. He was surfing off Kahana Beach, Maui, early in the morning when the Tiger shark attacked. Another surfer, Rodger Coombs, was paddling out when he heard cries for help. When Coombs reached McGinnis, he could see his right leg had been severely bitten by a shark. He got off his board and tried to push McGinnis to shore on his longboard. Others rushed to help when the two men reached land, but McGinnis collapsed. Despite CPR efforts by Coombs and the others, McGinnis stopped breathing a short time later.

  Retired police officer Bryan Lamy said he and others had tried to apply pressure to stop McGinnis's bleeding, but their efforts proved unsuccessful: 'His leg was severely lacerated. There was a lot of blood in the water and the victim was very pale.' Maui police captain Charles Hirata confirmed the injuries were consistent with a shark attack, saying: 'There was substantial arterial damage and a lot of blood loss.' The wound was about 35.5cm (14in) long. McGinnis had been surfing in murky water with low visibility: the islands had experienced abnormally high rainfall during the previous month, which had deposited large amounts of silt into the water. In addition, the late-season large north swell had churned the ocean water. He had just missed catching a wave and had turned to paddle out for another one when he was attacked, witnesses told police.

Shark expert Randy Honebrink insisted attacks were 'an extremely rare event and there's no pattern on how it happens.' He suggested ocean users should avoid murky water. Jeremy Franks, director of guest activities at the nearby Noelani Condominium Resort, observed: 'It's a local hang-out; it isn't known as a shark hang-out. It is the first time I think it has ever happened.'

In the last decade, Hawaii has seen a large number of shark attacks, which although not fatal, have often resulted in horrific injuries to the victims:

- On 12 November 2011, Jerônimo Pereira da Paz, 35, sustained injuries to both legs after being bitten by a shark while surfing off Del Chifre beach at Olinda. He said: 'I remember being very nervous. He bit my leg and let go.' Pereira da Paz swam to shore and was taken to the Hospital Miguel Arraes in Paulista, where he underwent surgery (he had ignored a surfing ban put in place because of possible shark attacks in the area). The month before, another surfer escaped with only a nasty scare and a bite to his surfboard when attacked by a Tiger shark at Nimitz Beach, Oahu.
- Surfer Theresa Fernandez had a lucky escape off Lyman's Beach on 25 May 2011. It was believed that a 3.3m (10ft) Tiger shark – the most common predator off Hawaii – was the species that had bitten the back of her surfboard. At first, Fernandez thought she had been bumped by a turtle but then

realised she was the victim of a shark attack when it tried to drag her under. 'I felt my board kind of get lurched forward and grabbed back and under, and I said, "Oh, this is not a turtle!"' she said. Luckily, she managed to swim to safety.

Later that afternoon Department of Land and Natural Resources officials re-posted shark warning signs. Although the beach remained open, the authorities strongly urged people to stay out of the water until they could determine the area was free of sharks for this was the second attack in four days. 'Given the timing and the location, it's likely to be the same shark but you don't know for sure. We do get more sightings around this time of year,' said Chris Stelfox, lifeguard water safety captain.

The incident did not deter Ms Fernandez from going in the water again – once she got herself a new surfboard. She said: 'The only thing that's stopping me from surfing is the fact that I need a board, so I'm trying to figure out how to make that happen.'

- On 21 May 2011, grandmother Alaina DeBina and her three-year-old grandson were on a paddleboard looking at turtles off Lyman's Beach, Kailua-Kona, when a 3.04m (10ft) Tiger shark bumped them. They were knocked into the water and a bite-sized mark was left on the board. Miraculously, both escaped injury. Recalled Mrs DeBina: 'Initially I thought, "Wow, I just hit a turtle!" It knocked me to the left of my board; it knocked my grandson to the right of my board. I landed on the shark and I was touching it with my hand trying to figure out what it was. It came up and took a taste of my board – it basically chomped on my board. Then it turned around and it was flashing its tail at me. That's when I was screaming for my husband, "Shark, shark!" – I was petrified at that point.' Mrs DeBina said she kicked at the shark and paddled her grandson back to shore.

Randy Honebrink, the local shark expert with the Department of Land and Natural Resources, conferred with Dr. George Burgess, director of the International Shark Attack File in Florida. The two agreed it was a 'huge coincidence' that the two women had been attacked in the same area within four days as there had never previously been a recorded shark attack in that location. Mrs DeBina said: 'I hope there was only one because it is a real scary thing to think there are two aggressive sharks out here on the same bay trying to eat surfboards, but I don't really know if we'll ever know the answer to that question.' Nevertheless, the brave grandmother added: 'Don't be afraid to get back in the water because number one, this is their home not ours.'

- Vaun Stover-French, 15, was bitten on the ankle by a shark while he was bodyboarding near Kahului Harbour on 24 December 2010. He said: 'I was just freaking out, just trying to get out of the water. I didn't know what was going on; it surprised me. I felt it grab my leg. I looked back and I saw the shark on my leg. I felt it on my leg, so I pulled back and it cut on my leg and it bit on my ankle.' He described his heel and foot as left looking like 'hamburger meat.' In fact, the gaping wounds needed 60 staples. Vaun's father Randy commented: 'He was pretty fortunate – he's got some good scars to show.'

- On 19 April 2010, Jim Rawlinson, 68, actually straddled a 4m (14ft) Tiger shark when it attacked him as he surfed at Hanalei Bay. The shark took a bite out of his board, sending Rawlinson high into the air – only to land on its back! Rawlinson grabbed its fin, pulled himself onto its back and held on tight: 'I was on the shark's back for anywhere from about five to ten seconds. It was so strange that everything was so slow and yet again so fast.'

  The inadvertent surprise counter-attack gave him time to undo the board leash attached to his ankle and make his escape. Local resident Leslie McTaggart who witnessed the astonishing scene said the water 'boiled' as the shark swam by and bit Rawlinson's board, adding: 'The shark was spitting pieces of the board out right under me – the guy could've died.' Marine biologist Terry Lilley said the attack was likely to be another one of mistaken identity and the shark had probably thought he was snapping at a turtle – 'The problem is we all look like turtles,' he said.
- Scott Henrich, 54, came off worse even though he punched back at a shark that attacked as he surfed at Kalama Bowls, Kihei, Maui, on 19 October 2009. He had just paddled out about 300 yards and sat up on his board when the shark suddenly emerged from the water and clamped onto his leg. Henrich recounted

how he had punched the animal's snout twice and it released its grip: 'Its head came up and he was on me – I never saw it coming. I thought, "I gotta get him off my leg," so I pounded him. I was just hoping he wasn't going to chomp all the way through. There was blood everywhere: I looked at the top gash and saw white meat, and I knew it was bad. I was worried there might be other sharks around, and I knew I had to get it up and get out of the water quickly.'

He managed to hobble to the nearby shopping centre, 'screaming the whole way' whenever he put pressure on his mangled right leg. After flagging down a motorist to call 911, Henrich sat and waited for an ambulance while several bystanders offered their T-shirts to help stem the blood flow. Eventually he was taken to Maui Memorial Medical Center in Wailuku, where he was given around 100 stitches to his wound and released. He said doctors estimated the bite radius of the shark at 61cm (24in). There was a big piece bitten out of the board and out of Henrich's shorts.

The state Department of Land and Natural Resources closed beaches along the South Maui coast from Lípoa Street to Kamaole Beach Park, during which time the waters were scoured for sharks. An extremely forgiving Henrich said: 'Tell all the surfers that I'm sorry they couldn't go out today because they closed the beaches,' before adding that the attack (which followed several previous shark encounters) would not put him off surfing: 'You are out in the food chain, and I was the first one out there and it was just my turn.' A Tiger shark was suspected of the attack.

- On 6 August 2009, a victim identified only as 'D. Crawford' escaped injury while surfing at Kau Kawa, Hawaii. Like many other surfers before him, his board was left with a large bite in it, consistent with the teeth of a Tiger shark.
- Mike Spalding, 61, was attacked by a shark believed to be a Cookiecutter (a species of small Dogfish shark), while attempting to swim the 48km (30-mile) Alenuihaha Channel from the Big Island to Maui, on 16 March 2009. A chunk was taken out of his left calf. Cookiecutter sharks, which are normally less than a metre long, are renowned for taking melon-sized balls of flesh from their prey. Spalding, who was inducted into the Hawaii Swimming Hall of Fame in 2008 for his seven successful channel swims between Hawaiian Islands, said the one he was attempting when he was attacked was his only challenge left – 'and the hardest one,' he added without irony.

He was taken to hospital, where he underwent a skin graft. Dr. Tim Tricas, professor of zoology at the University of Hawaii at Manoa, said he couldn't be certain Spalding had been bitten by a Cookiecutter shark but the description of his wounds was 'consistent' with a bite from that species.

Tricas said these sharks have very sharp teeth and usually prey upon pelagic fish (those that live near the surface of oceans and lakes) such as tuna, or porpoises. The creatures spend much of their time during the day in deep water but at night they come near the surface to feed, he added. They wound, but don't kill their prey. Spalding said the incident had not put him off his water challenges: 'Before long, I'll be dancing. I'll be back in the hunt, back trying to train for the channel again. I'm looking forward to the next time I get out there and finish this channel.'

Todd Murashige's encounter with a shark was literally life changing with the competitive surfer so grateful to be spared death that he turned to God. Murashige, 40, was surfing at Kahana Bay, Kaaawa, on 9 September 2008 when the 3.6m (12ft) Tiger shark pounced. Although he lost a lot of blood, he survived because no major artery was injured. He described the attack as being 'the perfect scenario for me. If I got banged by a car or hurt working, I wouldn't have got the message. Before it was like surf, myself, self-indulgence, not my family but it was something about being basically alone in the ocean that helped me get it and hear the wake-up call. To me, it's God's hand.'

After the attack, friends rallied round to raise funds for Murashige as he could no longer continue his work as a tile-cutter. Richard Peralta, who organised a fund-raising musical event, said: 'In Murashige's life, surfing used to come above all else – I hate to say it, even his family. We called him the samurai because he was stubborn and bull-headed. His eyes are more wide open now and he sees life in a bigger way.'

Murashige's wife Heather, mother of their two children

(Tyler, 10, and Tiffany, 8) said: 'It's a total transformation for him – he's been touched with something beyond what we can understand. He's embraced it.'

- On 26 July 2008, a swimmer identified only as 'U. Mataafa' sustained injuries to his left calf in the waters off Honokowai. Maui District police said his injuries were minor and that he drove himself to the Maui Memorial Medical Center emergency room and had told personnel he was bitten by a 0.6–0.9m (2–3ft) reef shark. Police chief Randy Awo said Mataafa reported he had been in shallow but murky water when he saw a school of weke fish, including one that was injured. He then felt a bump and realised he had been bitten by the shark.
- On 10 December 2007, Tino Ramirez showed his bitten surfboard to photographers after being attacked by a Tiger shark, off the north shore of Hawaii. The shark had become entangled in the board's leash. Said Ramirez: 'It thrashed about and pulled the surfboard backward and forward in an effort to break free. At one point the board was standing up in the water with its tail down as the shark tried to shake itself loose. I felt helpless. I slid off my board to try to get more control over my board. I wouldn't let go and it kept thrashing around.'

Eventually the shark bit the board, severed the leash and broke free. Ramirez added of his lucky escape: 'My feeling is it bumped me and figured I wasn't food, and it got tangled in my leash as it was going away. I think it was just a big mistake on the shark's part. I was in the water with it – if it had wanted to eat me, it could have done that right there.'

- California man Aaron Finley, 32, was swimming off Wailea on Maui's south shore and in front of the Four Seasons Resort on 29 October 2007 when he felt the jaws of a shark around his calf. Obviously not finding him to his liking, the unidentified shark then swam off. Finley recounted: 'I felt something hit my leg really hard and turned and looked, and I saw this big grey thing turn and swim away, and realised I'd just been bit. I was worried he was going to come back for more, but not too worried – I was just mainly trying to get back to shallow waters, where he was less likely to come back.'

  The shark left a deep gash in his left calf and a smaller wound in his thigh. Plastic surgeon Dr. Peter Galpin observed: 'The uniform characteristic is what I call "catch and release". Sharks bite with a force of about 6 tons per square inch so if this shark really wanted him, it would have had him.'
- On 28 August 2007, bodyboarder Joshua Sumait (15) escaped with only a 10cm (4in) gash to his right heel during a close encounter with a Tiger shark that he kicked away in waters near a restaurant at Kaaawa. He was about 400 yards from the shore when a friend saw the shark and shouted a warning. Sumait said: 'My fin was in its mouth when I turned around. If it had bitten, my whole foot would have been gone. I thought I was going to die right when everything was happening; I thought it was going to come back for more. I'm glad it didn't!'

- Snorkeller Harvey Miller, 36, tried to fight off a shark but sustained serious leg wounds in the attack off Bellows Beach, Oahu, on 18 July 2007. One of his rescuers, Don Ewing (who was working in a civil engineering team) said: 'The guy was talking the whole time, even while we were pulling him into shore. He was thanking us and saying, "I hope I don't lose my leg."'

  He added that another rescuer (known only as 'Ray') was the other man who bravely entered the waters to bring Miller back to shore: 'The victim could have been in very serious trouble if that guy hadn't got out there when he did. I held a tourniquet on the victim's leg above the knee for at least 15 minutes. From the knee on down, the guy's leg was pretty chewed up – I'd say there were four to five lacerations. The cuts were deep; there were places where it looked like they were to the bone.'

  People on the beach later reported seeing a shark attack at least two sea turtles off Bellows and Lanikai Beaches, killing one of the turtles. Witnesses described the shark as a 2.4m (8ft) Tiger shark, but authorities at the scene said this had not been confirmed.
- On 24 June 2007, Alicia Yamada escaped injury but had her surfboard bitten while paddling close to shore at Mokuleia, Oahu. The teeth marks were consistent with an attack from a regular predator, the Tiger shark.
- Peller Marion, 63, sustained severe foot and leg injuries when attacked as she snorkelled 22.8m (25

yards) off Keawakapu Beach, Maui, on 7 March 2007. Describing the attack, she said: 'It just was like that – it really frightened me. It looked like a big shark to me. All of a sudden I felt the strangest feeling; I felt something clench onto this foot and the first thing I saw was one of my new flippers pop off. And I know I swallowed some water and stuff, and then I just kind of pulled myself together and headed toward shore. I'm really happy I got away with my life – I really am!'

- On 5 January 2007, a shark bit a chunk out of a Rich Reed's surfboard at Major's Bay, Kauai. Reed, 24, managed to get safely back to the beach and later a piece of his board washed up on shore. It bore a semi-circular bite mark around 33cm (13in) wide and 15cm (6in) deep.
- Kyle Gruen, 29, paused to admire a few fish during a swim off Kihei's Kamaole Nalu Resort, Maui on 11 November 2006, when a large shark closed its jaws around his left thigh and hand. He recalled: 'It hit me from the side and I felt it clamp down with a lot of force. From then on, it was pretty much reflex. I was turning to get away and I looked at him: it seemed as big as I was and I've since heard it was anything from 6–12ft in length. I'm sure he was as scared of me as I was of him.'

The shark left five big puncture wounds, including a large tear thigh from just above the knee to a few centimetres away from the groin. It also injured Gruen's left hand in the same bite, severing tendons to leave his fingers flopping as he tried to swim for shore. 'I was pretty scared for my life at that point. I was in shock, it didn't hurt at all but I was afraid I was going to bleed out. I remember looking down at my leg and watching the little pools of water on the rock filling with blood – I could see the tendons in my hand,' he said.

Paramedics arrived within minutes and Gruen was taken to hospital for surgery. Incredibly, he was still able to carry out his duty as best man at a friend's wedding two days later, but had to leave early and admitted to having nightmares following the attack.

- On 31 May 2006, Ronald Deguilmo, 26, was bitten on his left arm while spearfishing at Kapaeloa, Oahu. 'I heard him screaming, "I got hit! I got bit by a shark!"' said his friend James Santiago, who was present at the time. 'It looked pretty bad: it was very bloody, it was deep.' Santiago said that as he and another friend swam Deguilmo to shore, he told them that he had whacked the shark with the butt of his spear gun: 'He told us that he hit it when the shark was tugging at his arm. When he hit it, it let him go.'

  The two men took turns, one supporting Deguilmo and the other swimming behind with a speargun, keeping a lookout in case the shark returned. Later, Deguilmo was told he would probably not regain all the feelings in his fingers.
- A man identified only as 'A. Balmaceda' was bitten on the left knee while spearfishing close to Club Lanai, Lanai on 24 May 2006. He had prodded the Grey Reef shark with his spear.
- On 23 March 2006, Elizabeth Dunn, 28, suffered puncture wounds to her left calf while surfing at

Kapaeloa, Oahu. She said: 'It was after he bit me, and the water calmed down a little bit, that I was really, like, screaming my head off. The doctor said that based on the way the bite marks are, that it was really big – that it could have swallowed me. So I just feel really, really lucky.'

- Nicolette Raleigh, 15, had a narrow escape on 29 April 2006 when swimming at Big Beach in south Maui in a celebration gathering for her 17-year-old boyfriend, Shane Wilds. She recalled: 'Something came along and knocked Shane down, then the same thing came over to me and I felt something shaking my leg. I thought it was my friend trying to knock me over. I saw the shark shaking its head backwards and forwards and I just started to scream. I was way too scared. When I got out, I looked at my leg – it was very ugly. I didn't believe it; I still can't believe it.'

  Nicolette suffered a long deep wound to her leg and was told by doctors it would be nearly eight months before she fully recovered. Said the brave teenager: 'I just feel I'm really lucky – I really didn't think I was going to survive because there was so much pain.'
- On 21 December 2005, swimmer Jonathan Genant, 29, was attacked by a Tiger shark at Keawakpu Beach, Maui. His left hand was bitten.
- Clayton Sado, 22, escaped with only a damaged surfboard after a Tiger shark lunged at him at Honokowai, Maui on 13 October 2005.
- On 18 June 2005, Brad Grissom, 49, punched at a Tiger shark as he was swimming off Maui and

managed to see off his attacker. He said: 'He was two feet away from my face when I threw the punch – it was pretty amazing. Then I just bolted straight to shore.'

- A man identified only as 'J. Bailey' sat in fear as a Tiger shark took a bite out of his kayak off North Kilhei, Maui on 14 May 2005.
- On 2 May 2005, Scott Hoyt, 47, watched as a Tiger shark bit his surfboard off Noreiga's, Maui. Miraculously, he escaped injury.
- Greg Long narrowly missed death when knocked off his surfboard by a Tiger shark at Rocky Point, Oahu on 16 February 2005. The shark bit at his board before retreating.
- On 9 October 2004, Davy Sanada, 34, had his left shoulder bitten by a Tiger shark as he was spearfishing off Mololkai. Sanada – a Pearl Harbour pipefitter – was freediving alone in shallow water outside Kupeke Fishpond when he was attacked.

He was swimming back to shore when the shark 'came out of nowhere.' Sanada said: 'Something had me and then I started flailing away at it. I lost a lot of blood and was getting dizzy. It was an ordeal, but I was yelling for help and nobody was responding.' He managed to hit the shark with his speargun, pull his wetsuit over his shoulder to staunch the flow of blood and somehow made it back to shore. The shark was estimated to be about 3.6m (12ft) long and was attracted by the bag of fish Sanada had caught. Sanada had no ill feeling for his attacker, despite such a severe injury, and said: 'If I see a shark, I give them all the room in the world. We respect each other. You've got to realise that's their domain and give them all the respect.'

- Bruce Orth, 51, escaped injury while surfing off Kalihwai Beach, Kauai, on 16 March 2004. The Tiger shark, described as being up to 3m (10ft) long, retreated after taking a bite of the surfboard.
- On 30 December 2003, top amateur surfer Bethany Hamilton, 13, lost an arm in an attack at the popular surfing spot Tunnels at Makua Beach, off Kauai's north shore. She had been surfing with her best friend and fellow competitor Alana Blanchard, Alana's father Holt and brother Byron. The attack happened around 7.30am as a shark, described as between 3–5m (10–15ft) lunged at the girl and bit off her left arm, just below the shoulder.

  'Nobody saw the shark. She paddled over to her friends after the attack with just one arm,' said Bethany's 17-year-old brother Noah. 'The biggest news is that she never cried once. Losing her arm will change a lot for her, but she never cried. The doctor was amazed at how well she was holding up. She told one of her friends that she's glad this thing happened to her, "Because now I can tell the whole world about God."'

It was an amazing declaration, not only for one so young but because Bethany was the second-ranked surfer from the Hawaii area, had been highly-placed in numerous competitions and was a former state champion. 'We were just trying to recruit her for the world team. She's probably, right now, the strongest waterwoman in Hawaii. Her strength in big waves, her paddling ability, her water knowledge, she's the best. And her character is such that she will overcome this,' said head coach Rainos Hayes. Noah added: 'She might have a hard time, but there are quite a few surfers who've surfed with one arm.'

- Shawn Farden, 16, was badly bitten on his left foot by a Tiger shark on 28 August 2002 as he surfed at Kewalo Basin Channel, Oahu. National Marine Fisheries Service biologist John Naughton said he inspected the tooth imprints on the surfboard and concluded it was a large Tiger shark, saying: 'It was a classic hit-and-run attack. It took the tail right off the board. On biting the board, the shark likely concluded it was not something he wanted a second taste of.'
- On 11 June 2002, a C. Levin had a lucky escape off Anini Beach, Kauai, when only his surfboard was bitten.
- Tommy Holmes, 35, sustained several lacerations to his buttocks by a Tiger shark as he snorkelled close to shore in shallow water at Olowalu, Maui on 1 January 2002. He was with friends, including girlfriend Monica Boggs, who were watching a large group of turtles when the shark attacked. Holmes said: 'It was amazing – we were just watching them for about 10 minutes when Monica spotted the shark about 25ft [7.6m] away and grabbed my hand. I put my mask back in the water to see where he was and he was around 4ft [1.2m] away. I saw his open mouth and teeth, and a very big head.'

To protect his limbs, Holmes curled up into a

ball and the shark latched onto his buttocks, then quickly released. Before it swam off, Holmes managed to punch its snout as it lingered near the surface. Monica said: 'It was swimming right at us at an alarming speed. It didn't look curious – it looked like it knew what it wanted. I thought we were going to die.' Holmes said he was 'ecstatic' to be alive. Randy Honebrink of the local Shark Task Force commented: 'He was very lucky, that's for sure. Usually Tigers would be expected to do a little more damage than that.'

- On 14 November 2001, M. Schweitzer escaped with only damage to his surfboard when a shark attacked at Kapalua, Maui.
- A victim identified only as G. Dano was bitten on the hand by what was thought to have been a Whitetip Reef shark at Ewa Beach, Oahu on 11 April 2001.
- On 23 March 2001, Mike Mendez, 27, was bitten on his left hand and saw the front of his bodyboard disappear into a shark's jaws at Sandy Beach, Oahu. The attacker was thought to be a Reef shark. Said Mendez of the attack: 'I thought it was one of my friends playing around at first, then I felt three tugs on my board and noticed that a piece of board was floating in the water and that my hand was bleeding.'
- Henrietta Musselwhite, 56, was attacked by a Tiger shark believed to be up to 3m (9ft 10in) long as she swam and snorkelled off Olowalu, west Hawaii on 18 October 2000. The right side of her back and torso were lacerated. Kayaker Ron Bass was paddling nearby at the time. He recalled: 'There was a bunch

of Whitewater around her. I saw this grey thing come out of the water three times – she yelled for help really quick when it happened.'

Bass added that he put Mrs Musselwhite on his kayak and brought her to shore. He said the two wounds to her back were 'more like slashes – about 3in wide. One was 6in long; the other, about 4in. The wounds were really bad. If she had swum back with those lacerations, she probably would have bled to death.' The attack took place less than a mile from where West Maui resident Marti Morrell was killed by a shark while swimming near her beachfront home in 1991.

- Jean Alain Goenvec, 53, had his left calf badly bitten by a large Tiger shark as he was sitting on his windsurfing board off Kanaha Beach, Maui on 15 August 2000. Water safety officer Joe Perez launched his rescue ski and sped out to Goenvec. Perez said the victim was fully conscious and had used one of his sail lines as a tourniquet in an attempt to stop the flow of blood from his massive wound. Goenvec was able to cling to the sled attached to the back of the rescue ski and was towed ashore. He was taken by ambulance to Maui Memorial Medical Center, where he was initially listed in a critical condition but then treated for his injuries.

  Maui County Chief of Aquatics Marian Feenstra said the incident showed the importance of equipping beach lifeguards with rescue skis and sleds so they can respond quickly to emergencies. It would have taken much longer to bring Goenvec to safety if officials had to rely on the Department of

Fire Control to launch its rescue boat from Kahului Harbor, she added.

ONE IN THE EYE FOR A SHARK

On 24 March 2002, when Hokuano Aki, 17, was attacked as he bodyboarded off the coast of Hawaii, his instincts for survival quickly set in. Despite being twice dragged under the water, he jabbed the shark in the eye. Sadly he still lost his left foot in the attack at Brennecke Beach. 'He said he was being thrashed around underwater. He fought back and was able to grab the eye, and the shark released him,' said Fire Captain Mike Layosa. 'It looked like he got hit just below the knee and that the shark stripped much of the flesh from the lower left leg before removing the foot.'

A witness recounted how he had seen Aki enter the water shortly before noon and the shark attacked about five minutes later. At the time, the water had been muddy with runoff from heavy rains. The witness – the husband of a registered nurse who later helped stop the boy's bleeding – said he saw a fin about 30cm (12in) long, the teenager was pulled under, came up yelling and was pulled under again. The second time he surfaced, he swam to shore, kicking with the fin on his remaining foot. Aki later told his shark survival story from a hospital bed:

> I opened my eyes and I could see the shark. It was just tossing me all over the place. I remember hearing... I thought I heard my leg break. I heard the bones snap. I remember I tried to open the mouth and tried to get it off of me; that didn't work either. I just grabbed the shark's eye and ripped it out and then it let me go. I had

> a look at my leg and I just noticed the skin was all torn up and all my flesh was just torn up. I didn't really notice my foot was gone until I was in the ambulance. I was with my brother at the time, and he told me foot wasn't there and I was just in shock – I couldn't believe it. I still can't believe it. I'd like to thank the nurse who was at the beach: she saved my life.

It was a trauma nurse visiting from Colorado who was at the beach when Aki came ashore and who was credited with saving his life. When a lifeguard from Poipu County Beach Park arrived, she had already stopped the bleeding. Beaches along the south shore were closed and there was a report of sharks sighted off Lawai Beach Resort. The Kauai Fire Department issued warnings against going into the ocean during periods of heavy runoff because that's when sharks come close to shore scavenging for food.

One visitor to Aki's hospital bed was Michael Coots, who lost his lower right leg in a shark attack while bodyboarding at Kauai's Majors Bay in 1997. Coots was back in the water six weeks later. He said his visit was as much for Aki's parents, Harmon and Kalea Aki, to let them know their son would recover. 'The hardest part, honestly, was sitting on the beach, watching all my friends surfing while I waited for the stitches to heal so I could surf again,' he admitted.

'Jabbing the shark was precisely the right thing to do,' said National Marine Fisheries Service biologist John Naughton. 'They generally release where the guy hits or fights, especially around the sensitive areas of the gills or the eye.' Although neither Hokuano nor witnesses could identify the shark, Naughton described attacks such as this as 'classic Tiger shark attacks.' 'Often,' he said, 'a Tiger shark will attack a human in murky water, do a great deal of shaking and leave. If a turtle were attacked, the shark would return for more.'

The attack prompted more warnings about the dangers of swimming in dirty coastal waters, during and after periods of heavy rainfall. Naughton said sharks have a feeding advantage in muddy water. State aquatic biologist Don Heacock said that without accurate information there was no way of knowing for certain what kind of shark it was, but he agreed that a Tiger shark was a likely suspect: 'These animals are the wolves of the sea and they'll put themselves in a position that's to their advantage. In muddy water, they have a great advantage because they can detect smells and motion but they'll attack in crystal-clear water, too.'

Hokuano underwent surgery. His father Harmon said: 'He's a strong boy. We're thankful – it could have been even worse. We still have him here with us and that's good. No doubt it's going to be a long journey.'

The attack prompted warnings about the dangers of swimming in dirty coastal waters during and after periods of heavy rainfall. Don Heacock said the Waikomo Stream is not far away, flowing dirty water into the ocean. Southerly winds had kept the murky water close to shore. Meanwhile, Naughton was of the view that sharks have a feeding advantage in muddy water because their nonvisual senses are so acute: 'They may be more likely to come near shore when the water is dirty in hopes of finding one of their favourite food sources, green sea turtles.' He had recently visited the Po'ipu area and saw many turtles in the near-shore waters.

But sharks often bite other things as well: the stomach contents of Tiger sharks fished off south Kauai during the late

1960s and early 1970s surprised researchers with the number of land mammals they found. Said Naughton: 'We got some amazing things out of the stomachs of those sharks – we found a horse's head in one of them. That's one of the reasons we're telling surfers and others to stay away from stream mouths when they're in flood.'

Nearly half of Hawaii's 113 shark attacks since 1882 have occurred off Oahu (37, with six fatalities) and Maui (37, with three fatalities). In recent years there have been 20 attacks off Kauai with no fatalities, 12 off the big island of Hawaii (one fatality), two off Molokai (no fatalities) and one off Johnson Island (no fatalities). In 1959, following a fatal attack, Hawaii launched a government-sponsored shark eradication programme, which was to last 10 years.

Some native Hawaiians call the Tiger shark '*aumakua*' (guardian spirit).

SHARK SPECIES OF HAWAII

There are around 40 species of shark in Hawaiian waters. The most commonly seen are:

- Tiger shark
- Whale shark
- Pygmy
- Sandbar
- Reef Whitetip
- Scalloped Hammerhead
- Grey Reef
- Galapagos
- Great White

NORTH PACIFIC NIGHTMARE

When marine scientist Kydd Pollock, 33, attempted to free a 2m (6.5ft) Reef shark entangled in a net on 19 December 2010, he was rewarded by a frenzied attack to his face and head. He was only spared more serious injury because of his diving mask. Pollock had been carrying out marine research in a lagoon at Palma Atoll in the North Pacific with other divers when he noticed the shark caught in a net the team were using to catch giant hump-headed Maori wrasse for tagging. The shark was released but headed straight for Pollock.

Describing the incident, his father Rick said: 'This 6–7ft [1.8–2.1m] Reef shark swam into one of the nets and got caught. My son was on the other net. They cut the shark out and once it got out of the net, it panicked and it made a beeline for the other net. It looked like the shark was going to swim right into the net again and get tangled, so Kydd grabbed the floatline and swam down to the bottom with it. The shark swam right past him, went through as he had hoped but spun around – and that's when the attacks started on his head. The first bite was on the back of his head. The second bite, which was potentially the worst, was the one that his mask took the full brunt of. It shattered the glass and twisted the mask into a pretzel, so I just can't imagine what sort of force went into that. And then it came back for a third time and grabbed him on the forehead and the top of the head.'

Not only was Pollock spared potentially fatal injury but also by an amazing stroke of luck, his girlfriend had enough medical knowledge to treat him. This probably saved great blood loss as the remote island's nearest hospital was up to a day's journey away. Said Rick: 'She was there on the island; she's a PhD, not a medical doctor. They didn't actually have a medical officer on there – she was the next best thing. She's the one that stitched him and sewed him, and stapled him up. I thought it was quite remarkable that she would be able to divorce herself of any sort of emotional entanglement and was able to deal with the issue as it stood. She shaved his head and did everything that was required, and I really take my hat off to her. Kydd was extremely lucky considering he was attacked on the head and the fact he was on such a remote place with few medical supplies and little prospect of getting to hospital.'

Palmyra is halfway between Hawaii and American Samoa.

SHARK SPECIES OF THE NORTH PACIFIC

Around 40 species of shark inhabit the North Pacific. These include:

- Broadnose Sevengill
- Salmon shark
- White shark
- Shortfin Mako
- Tiger shark
- Hammerhead (three varieties)
- Silvertip shark
- Bignose shark
- Copper shark
- Silky shark
- Galapagos shark
- Bull shark
- Blacktip shark
- Oceanic Whitetip
- Dusky shark
- Sandbar shark

- Whitenose shark
- Lemon shark
- Blue shark
- Nurse shark
- Goblin shark
- Ragged Tooth shark
- Cookiecutter shark
- Pacific Sharpnose shark
- Mallethead shark
- Scoophead shark
- Bonnethead
- Thresher shark (three varieties)
- Megamouth shark
- Pacific Angel shark
- Frilled shark
- Prickly shark
- Spiny Dogfish
- Combtooth Dogfish
- Pacific Sleeper shark
- Pygmy shark
- Horn shark
- Mexican Horn shark
- Whale shark
- Crocodile shark
- Basking shark
- Swell shark
- Catshark (four varieties)
- Smoothhound (four varieties)
- Leopard shark
- Soupfin shark

SHARK MAYHEM IN MEXICO

On 23 May 2008, surfer Osvaldo Mata Valdvinos, 21, was attacked alongside Troncones beach, Guerrero, Mexico. He died after the shark bit his left hand and broke one of his legs. 'Two witnesses, his friends who were swimming with him, told us they saw a 2m (6ft) shark attack him and pull him underwater,' a police spokeswoman said. Friends paddled Valdvinos back to shore, but he lost consciousness and died before medics arrived. Said one onlooker: 'He raised his hand and screamed, and that was when his hand and leg were bitten. He went under; he was under the water for a while and then he emerged bleeding.'

On 28 April 2008, San Francisco surfer Adrian Ruiz, 24, died from blood loss following a shark attack off the same beach. He was taken to the shore after being bitten on his right thigh and suffering a large wound. An official report described the wounds as 'reaching from the hip to the knee, exposing the femur.' The injuries were consistent with the teeth of a Great White.

Diver Ken Pitts, 45, got more than a close-up of sharks when he swam in an aquarium display tank at Albuquerque, New Mexico on 12 March 2005. A Sand Tiger shark took offence at the intrusion and bit him on the forearm to leave two nasty puncture marks after beast and diver collided. The mishap was recorded as a 'provoked incident'.

CHAPTER 10

SHARK ATTACK SURVIVORS BECOME THEIR SAVIOURS

'MAKE NO MISTAKE ABOUT IT, SHARKS ARE KILLERS BUT I DON'T HAVE ANY REASON TO HATE THEM...'

In February 2004 Debbie Salamone survived a shark attack in Florida's waters. She had been wading in the surf with her husband, Craig Wickham, at Canaveral National Seashore when a storm rolled in. The couple were heading back to shore when Debbie said she saw a 'large fish' leap out of the water – 'Immediately it clamped on to my heel and I tried to shake it off; it was just this frenzy of kicking and me screaming.' She managed to shake the shark off, but then it bit her again. Later, 43-year-old Debbie gave an account of the episode to the *Sunday Telegraph Magazine*: 'A storm was approaching so I began heading ashore. Then a big fish jumped out of the water some 10 feet [3 metres] away and I grew alarmed. As soon as that thought formed in my head, the shark got hold of my ankle, biting down hard on my Achilles tendon. The pain was intense and terror shot through my body as I kicked and screamed. My arms were outstretched and I was screaming, "It's got me. It's got me! It's got me! It's got me!" Then I was dragged ashore. That's when I looked down and saw a bloody pulp where the shark had severed my Achilles and bit into the sole of my foot. All I could think of was, "Will I dance again?"'

She said the water was too murky to see the shark properly but she felt a slimy body go past her left foot. The attack lasted no more than eight seconds but her foot was shredded. Luckily, a local nurse who was walking on the beach tended to her injuries while her husband dialled 911. The nurse, Tecia Lucignani, wrapped the injured foot in a towel and applied pressure. 'She had seen me lying face down on the beach, blood draining into the sand,' said Debbie. 'But the tide was rising, relentlessly chasing us up the beach and forcing me to keep crawling awkwardly to dry land. When the rescue workers arrived, they cleaned my wound, then one of them flopped me over his shoulders – fireman style – and the other grabbed my legs. Thus, we exited from the beach to the backdrop of crashing thunder and lightning so typical on a Florida summer afternoon. At the hospital I remember screaming at the doctors "I'm a ballroom dancer!" hoping to get them to do a good job patching me up, which they did.'

It took years of acupuncture and laser therapy for Debbie to return to her former health, though. The attack had a double-lasting effect on her: not only did she have enduring injuries but after an understandable initial hatred of sharks, she became a leading light in worldwide shark conservation. Despite the ordeal, she had had something of a vision: 'I began to see this as a test of my commitment to environmental conservation,' she said. As a result, she began working with the Pew Environment Group, which runs a worldwide shark conservation campaign. She also formed a group: Shark Attack Survivors.

In 2009, Debbie wondered if others who had been through the same traumas would nevertheless be prepared to join in the fight on behalf of sharks and so she tracked down as many as possible. Despite often-severe injuries – including the loss of arms and legs – many agreed to come on board the campaign:

- Chuck Anderson, 54, a high school athletics coach, was attacked off Mobile, Alabama, on 9 June 2000, while training for a triathlon. A 2m (7ft) Bull shark hit him from underneath, biting off four of his fingers before shearing off his right arm at the elbow. Even ten years on, Anderson could recall the attack vividly.

  'It was a Friday morning. A big group of us were meeting at the end of Highway 59. One of the group said the water was too rough for him, so he stayed on shore. Myself and the two others were all strong swimmers so I suggested we go ahead and test the water to determine if the conditions were safe for the others. About two minutes into the swim, I remember looking at my watch. It was 6.38am. I swam two more strokes when I felt something hit me from below. I didn't know what it was, but I knew it was big. It literally lifted me up out of the water. I panicked and just started hollering for everyone to get out of the water. When I came to my senses, I began to look around and didn't see anything so I put my face down in the water; that's when I saw it. I put my hands out instinctively to fend him off.

  'His first bite took all of the fingers and my thumb off my right hand; I don't remember feeling any pain. I pulled my hands back and tried to hold my right arm up out of the water to keep the blood from going into the water. It was not very successful.'

The shark lunged at Anderson's stomach but for some reason then let go: 'I would always be sure I had the shark by his nose to keep him from tearing me up because those teeth, they were just like machines coming after you so I was keeping him away by holding his nose when he would charge me. Then every chance I would get is when I would hit him. Whenever the shark would break away, he would make his turn; I guess he would do a circle. I would swim as fast as I could and I'm just assuming it's probably a five- to six-second burst I would get. I would always turn back to see if he was going to charge me again. Each time I would turn back, he was there. I mean, he was there – he was charging each time. The next strike, I actually saw his fin coming through the water at me; that's when he grabbed me by the arm and went into a feeding frenzy, throwing me around like a rag doll. The shark practically had my entire arm in its mouth; I was having to hold it back to keep my facc away from his mouth. That's when I had a conversation with God; I said if he helped me, I promised I would do certain things. I told him that if I survived, I would wake up every morning thanking him for the opportunity to live another day, and I would go to bed each night thanking him for the blessings of that day.

'At that moment the shark began to push me toward the beach. I was able to get my feet underneath me and stand up. Then I tried to rip my arm out of the shark's mouth. My arm from just below the elbow was stripped to the bone (they call it de-gloving). As soon as I was free of the shark, I began to run to the beach as fast as I could. A young man on the beach took off his shirt and I helped him apply it as a tourniquet and I laid there on the beach until paramedics arrived.

'I know I am very fortunate to have survived. If the shark had clamped down on my stomach, or if he had turned and gone away from the beach, I would not have been here to enjoy these last 10 years since the accident.'

Incredibly, despite losing four-fifths of blood from his horrific injuries, Anderson did not lose consciousness. He was taken to South Baldwin Hospital in Foley. 'I made them keep me awake long enough to see my children and tell them that I love them. After that I don't remember anything for five days,' he said.

The skill of surgeon John Rodriguez-Feo and his team saved most of Anderson's arm. 'It really makes a big difference. I am able to hold onto things and even grab some things. I can do so much more than if I didn't have anything there below the elbow,' he said. After undergoing the emergency surgery, Anderson was at South Baldwin for two days before being transferred to another hospital, where he underwent several more operations. Amazingly, he still goes swimming – but in a local swimming pool. To this day, he strongly believes it was the power of prayer that got him through the attack. He said: 'I really didn't know how close to death I was. I am definitely blessed to be here.' Even more amazingly, he joined the Shark Saviour Group: 'I know a lot of people will think it's strange that a shark attack victim would speak out against cruelty to sharks. It is a cause I truly believe in. This practice of shark finning is responsible for the slaughter of large numbers of sharks in a very inhumane way. It has gotten a lot of attention that shark victims such as myself have come out in support of this cause and I want to do what I can to spread their message. Make no mistake about it, sharks are killers but I don't have any reason to hate them.'

- Mike Coots, 30, a keen surfer from the Hawaiian island of Kauai, was 24 when he was attacked by what he believes was a Tiger shark on 29 October 1997. He lost a leg – 'I was paddling out and it was pretty quick. The shark came up from under me, and grabbed my leg and waved me around like a rag doll. It was over quicker than I thought, and I never even felt my leg come off.' His friends saved his life by using his surfboard leash as an emergency tourniquet to stem the bleeding before pulling him to the beach.

  'I remember flying down the road to the hospital lying down in the bed of the truck with blood just pouring out of my leg and down through the tailgate. I got attacked pretty early in the morning, around the same time the kids were going to school, and I remember passing all these cars full of parents and kids. As we passed them, they would look at me in the bed of the truck with my foot torn off and they would just pull over. I will never forget the look on their faces when they realised what was going on.'

In the following months Coots endured agonising corrective surgery and rehabilitation, but despite all that, he managed to carve out a career as a photographer. 'People often say, "You must be so mad at sharks, you must just hate sharks – do you want to kill sharks?" But I've come to realise that sharks predate the dinosaurs and we're in their environment. I have no hatred whatsoever for the animal and feel they really play an important part in our environment. I felt like I could open up doors with my story – and why not? It's a respect thing; it's not a fear thing. When I was contacted to help this cause it was a humbling experience because I do a little bit of work in Australia but to be able to come over here on a worldwide scale, go to the United Nations and work with these wonderful people, that's amazing. You try and make a difference worldwide.'

- Krishna Thompson, 44, a Wall Street banker, was celebrating his 10th wedding anniversary on 4 August 2001 when he went for an early swim at the Our Lucaya resort beach on Grand Bahama Island:

- 'I had swum out about 20ft [6m] and was treading water, looking out toward the ocean when I saw the shark fin speeding straight for me. I tried throwing my body toward the shore to get out of his way – I'm a quick guy. But the shark hit my right leg. Then he caught my left leg and I heard his teeth go into the bone, like in a cartoon. He towed me out into the ocean; he just kept going, and I was thinking, "I can't believe there's a shark on my leg." The water was frigid. I think that's why I felt pressure but no pain. I thought about my wife and how we didn't even have children yet. I thought: "I'm going to die." I remember saying aloud, "Oh God, get me out of this!"

'Suddenly, the shark took me under. I remember the swirling, like when water goes down a drain. The daylight disappeared into that little hole. That's when he began violently shaking me like a rag doll. It was dark. I just tried to imagine where its mouth was, felt around and pulled it open. I got free – I couldn't believe it. The grip was awesome (later, you could see the teeth marks on my bones). Then I gave the shark two quick blows to the nose – one, two – and it just swam away. All around me the water was red. As soon as I was clear of that shark, the first thing I did was look at my leg. All I saw was bone, no skin, no arteries; I was going out to sea. I thought, "Man, they're going to amputate!"

I just started to swim. I have no idea of how far it was to shore. Once I reached shallow water, I started hopping toward the beach. I tried to yell but it wouldn't come out. Finally, I reached deep down and screamed. That's when people came running. I wasn't feeling pain in the leg – I think all my nerves were gone. I couldn't move my good leg, arms or tongue. I kept fading, thinking it's a dream, and then I'd be like, "No, it's real. I can hear them

> working on me." My heart was pounding. Then it slowed and I thought my body was shutting down. The next thing I remember I woke up in a Miami hospital, happy because I was alive.'

Thompson had lost so much blood that he appeared clinically dead on the operating table. 'I remember feeling I was going to the other side. The surgeons were washing up when "Pop!" – a monitor sounded and they rushed back to the table and managed to save me. I didn't look to see the leg was amputated, I just knew: you sense it. Someone asked if there was anything they could get me. I said, "How about a leg?"

A month later, he got his leg: a computerised limb called a C-leg with a microprocessor in the knee that adjusts the way the leg flexes. Learning to live with the artificial limb was still a painful challenge, though. Thompson said: 'At first putting weight on my residual limb hurt like hell – sharp pain ripping your skin apart, like the sutures were stretching. The pain was so bad; I went to my first rehab class and left my leg in my room. They were like, "Go back and get your leg!" They were saying, "You're favouring it too much." I was like, "It hurts!" But they just pushed.

'The C-leg is the closest you can get to a real leg: it has a "skin". If I wore shorts, you wouldn't really know it's prosthesis. But I don't need to cover it up – I want people to see this is my leg, this is who I am. The company that makes the leg asks me to speak once in a while. I feel pleasure in helping someone else. They had me speak to this young guy. He lost his right leg above the knee; he was pretty down. I told him, "Don't try to run a marathon tomorrow. Try to do a little more in small intervals." You can't even imagine what you can do if you take baby steps. My daughter Indira was born on 26 September 2002. To go from thinking you're going to die and not have children to watching your child born in front of you, it's the greatest thing on earth. When you've been in the jaws of a shark and then you think about your worst day, there's no comparison. I thank my lucky stars I'm alive.'

Amazingly, Thompson said he wanted to save the sharks as well: 'I am blessed to be here, but the shark has a right to survive, too.'

• Paul de Gelder – a diver for the Australian Navy – was attacked by a bull shark on 11 February 2009 in Sydney Harbour. First, the shark attacked him and then de Gelder hit the shark. He said:

> 'I felt an almighty whack on the leg. I didn't think too much of it at first – it didn't hurt. Half a second later I turned over, looked down to check my leg and saw the huge grey head of a Bull shark, one of nature's most aggressive man-eaters. What's more, I could see the upper row of its teeth across my leg. Its lip was pulled back and its mouth looked enormous. We must have stared at each other for about three seconds, but as soon as I recovered from the shock, I started fighting for my life. I couldn't seem to move my arm; it was pinned down by my side. I hadn't realised my hand was also in its mouth. I tried to stab it in the eyeball with my other hand. I tried to push its nose, but my hand just slid off, like pushing a slippery concrete wall. I pulled back my left arm and punched the shark as hard as I could on the nose.
>
> It started shaking me, like a dog would a rag doll. The shark pulled me down under the water, continuing to

shake me. The second time I went under I could only see bubbles in front of my face; I no longer felt any pain. Being attacked by a shark is like getting hit in the leg with a plank of wood – you don't even feel the teeth go in. I think the adrenaline, the panic, probably puts a numb on the pain and you don't feel it. I couldn't do anything; I was totally helpless. Everything was quiet – there was just a deep silence.'

The shark disappeared... after virtually eating de Gelder alive. Said one witness: 'The attack occurred very quickly. The shark then disappeared very quickly – it was all over in a few seconds.' Doctors later amputated de Gelder's right forearm and leg. As he recalled in his book, *No Time for Fear*: 'I was just so overwhelmed. It hit me that this was now my life missing a leg, missing a hand. I was literally half the man I'd been and I wished that I'd died. At least then I wouldn't have had to go through this torture. I couldn't take it; it was too huge a concept that I'd live the rest of my life like this. How would I drive or dive, or do any of the crazy stuff I did before that I felt had made life worth living? There just didn't seem to be anything left for me. My life, as I knew it, seemed finished. It was a defeatist attitude and I hated myself for it. Previously, anything was possible and now everything seemed impossible. But eventually I ran out of tears and, you know what? Nothing had changed. I realised that I could lie there, crying and sooking [wallowing] until the end of my days, but I'd still be sans limbs when it was all over. I'd still be hurting, I'd still be struggling and, since suicide was out of the question, I'd still be alive. So I asked myself again, What the f*** am I going to do now? But after feeling sorry for myself for a bit, I was determined to just get on with it and make the best of it all.

'There are times, of course, when I curse the shark that, in just a few mad seconds, completely altered the course of my life. But then if it hadn't been a shark, it might have been that dog who mauled me when I was younger, any one of my many motorbike crashes, a stray bullet in the army or, with clearance divers now deployed in Afghanistan to dispose of Improvised Explosive Devices and unexploded ordnances, any incendiary.'

Three months after the attack this amazing shark survivor was back in the water. He then joined Pew Environment Group to try and protect the very species that had so dramatically changed his life and declared: 'Do we have the right to drive any animal to the brink of extinction before any action is taken? It's time to replace our fear of sharks with understanding and action, just as we have for lions and other top predators. Regardless of what an animal does according to its base instincts of survival, it has its place in our world. We have an obligation to protect and maintain the natural balance of our delicate ecosystems.'

• Achmat Hassiem, 22, experienced a shark attack at Mulzenberg Beach, South Africa, while swimming with his brother Tariq and a group of friends. After hearing his horrifying story, it's hard to believe he wanted to protect sharks:

> 'I caught something out of the corner of my eye, a black shadow in the water. I thought it was a seal or a dolphin, and then this fin broke the water. The shark was heading towards my brother. I screamed for the rubber duck [the life-saving boat] to get out to him. They didn't understand what I was shouting about – I was screaming, "Get Tariq,

get Tariq – he's in danger!" Then I started splashing, trying to distract the shark. The shadow changed direction. It was coming towards me and then the fin disappeared below the surface. I knew that when sharks attack, they like to come from the bottom up. I could just touch the bottom and I tried to make myself as big as possible. But the shark didn't attack – it bumped me and its body rolled along mine then its tailed whacked me.

'I was rocking, trying to keep my feet. I lost sight of the shark but I could see my brother further out; he was screaming something at me. Then I saw it coming: its mouth was open. All I thought was to try and get away from its mouth, so I put my hand out and tried to push myself on top of it. My hand was on the shark's head and I tried to get my right leg over it. I couldn't move my leg, and then I saw half of it was in the shark's mouth. It started violently shaking me; it was terrifying. I could feel my leg being torn apart, but there was no pain. I was in absolute shock. I was being attacked by a Great White! It shook me again and started trying to pull me under. The water was still quite shallow and sand was churning everywhere. The shark began to try and head towards deeper water and started picking up pace. I thought this was over, I was going to die. I remember the sound of the rubber duck's engine disappearing. I thought they had left me, but it was because I was getting dragged deeper. The ocean was becoming darker. I was still trying to get out of the shark's mouth. I was getting short of breath and I remember thinking, why don't I just let myself drown – that would be better than what the shark would do to me? Then I decided: no, fight.

'I hit the shark with my fists. A shark's body is coarse and it was like hitting sandpaper, a tank wrapped in sandpaper. Soon I had no skin on my knuckles but I had one good leg left and I was trying to kick the shark. Then it shook me again, twice, and so hard that on the second one there was this cracking sound, even under the water – my leg broke off. I swam towards the surface and sound started coming back. I stuck my hand out of the water and that's when I saw my brother in the rubber duck. He grabbed me, saying, "Don't worry, I've got you." I was hauled into the boat as the shark came back. It dwarfed the boat; it hit the underneath of the boat. My brother jumped on me to hold me. He closed my eyes so I couldn't see what had happened to my leg. Later he told me that there were perfectly cut triangles of flesh with bits of broken shinbones hanging out. But still I felt no pain.'

The boys reached the beach and waited for the air ambulance to arrive. Later that day, Achmat was operated on and woke in intensive care. 'The first thing I saw was my brother crying. That hit me hard. He saw I was awake and said, "Thank you." I said, "What for?" "Saving my life," he said. Then he said: "You know what happened?" "What do you mean?" "Look under the blanket," he said. I was scared to look. I looked and saw my leg was gone – that was the first moment I really knew what had happened.

'I'd always played sport; it was all I wanted to do. I went into this great depression. It was on my third day in hospital that the pain really kicked in. You feel like your leg is still there – I felt like I had cramps in my right foot, but of course I didn't have a right foot. It's the worst pain I ever went through. I was terrible.

I remember waking one night, there was blood everywhere and I was yelling for the nurses. I couldn't handle it.'

A passionate swimmer, he was later encouraged to join a paralympic swimming team: 'The first time I got back in the water was really difficult. The first time I went underwater, the fear kicked in – it took a couple of weeks to get brave enough to do it properly.' But there was still something he wanted to do: get back in the sea. 'That was the worst thing ever. I wanted to do the Robben Island swim and so went down with some open-water swimmers to start training. They said they would look after me. It was not long before I saw a shadow in the water – it was a rock, but it gave me such a fright. I sat on the shore and said, "I can't do this."'

Achmat said he felt like 'an injured seal' but he went on to become a successful competitive swimmer. He said: 'There are still nights where I sit down and thank God I survived the attack and have had all these experiences. People say sorry, but the rewards that have come are amazing: losing a leg is nothing compared to losing a brother.' He too joined the shark survivors group.

• Briton James Elliott, 26, came face to face with a shark while on a family holiday in Egypt's Sharm El Sheikh resort on 16 April 2010. During the frenzied attack on the second day of the holiday, Elliott, suffered severe injuries, including a severed Achilles tendon. He was saved by a passing boat. He said of the attack:

> 'I'll never forget the moment. I was treading water about 40ft [12m] from the shore with my dad, enjoying the sun. Suddenly something grabbed my left leg and I felt a massive pain. I kicked out with my right leg and managed to get the teeth off my leg. Then I looked down

and a 5ft shark was staring at me; I thought I was going to die. I could see the huge white teeth, which was the scariest moment of my life. It turned and swam away, but then dashed towards me so fast half of its body and the dorsal fin was out of the water. I can't say I'd thought twice about sharks before I was bitten on the leg by what I'm pretty sure was a Mako shark.'

He later said: 'I want to swim with Great Whites one day!'

- Yann Perras of Le Mans, France, had his leg severed by a shark while windsurfing off the coast of Venezuela on 11 April 2003. Speaking about the incident, he said: 'I've always loved being out on the water. As a wind surfer, the chance to travel to a resort on a small island in South America seemed like a dream come true. Yet, an attack by a shark soon after my arrival turned my family vacation into a nightmare. I'll never forget that day. It's hard to describe to people the exact sensation I felt looking down, seeing the unthinkable through the transparent waters as a shark seized my leg.'

  Like his fellow shark attack survivors, he turned the horrendous experience into a positive approach to life: finding the strength to go on and to support the shark protection campaign. He said: 'With the loss of my right foot, travel today is now much harder, whether near or far. However, I didn't hesitate for a minute when I was invited to New York City as part of an effort by a group of shark-attack survivors from around the world to urge the United Nations to better protect these remarkable, misunderstood kings of the deep. I don't blame all sharks for my injury, though one caused my impairment. After a period of emotional and physical recovery, I found that the incident had instead opened my eyes to the perilous state of the world's most captivating and important animals, which, if lost, could set off a cascade of harmful effects across the entire ocean ecosystem.

'Sharks are important predators that help maintain the balance of the ocean food chain. According to scientists, their disappearance is already having an impact on the health of our seas. I refuse to let my personal tragedy be compounded by another, far greater one. It's time for the international community to stop the senseless slaughter of sharks. Misplaced fear of these animals is no excuse for allowing fishing practices that are cruel and threaten the natural balance of life in our oceans.'

- While windsurfing off the coast of Venezuela in 2003, Vincent Motais de Narbonne's leg was severed by a shark. He said: 'Even if the movie *Jaws* has scared entire generations, we have to remember that it is only fiction. This animal is, like people, at the top of the food chain. We absolutely cannot accept fishing practices that menace the natural balance of the ocean environment.' (Shark attacks off the coast of Venezuela are few and far between, but on 22 January 2004, there were two attacks within an hour of each other. Rafael Gonzalez, 40, was bitten in the leg while fishing in Mochima National Park and then Stefarnt Moeller was bitten in the lower back and hand. Fishermen later killed the shark. Before that, the last recorded attack was in 1973.)

In July 2010, the group came together to launch their unusual lobbying campaign: attempting to persuade the American Government legislators to back a tough new US law to clamp down on the widespread killing of sharks. 'It's time to replace our fear of sharks with understanding and action, just as we have for lions and other top predators. Regardless of what an animal does according to its base instincts of survival, it has its place in our world. We have an obligation to protect and maintain the natural balance of our delicate ecosystems,' said shark survivor Debbie Salamone.

In September 2010, ocean conservation groups including Pew Environment Group reunited once more to urge countries to strengthen the protection of sharks worldwide. Although several voluntary national and international agreements exist that set guidelines for the management of shark populations, these have been largely ineffective in stemming shark overfishing. The survivors-turned-advocates are calling for an end to the fishing of sharks threatened, or near threatened, with extinction, to stop the practice of finning and for better managed fisheries to ensure long-term sustainability.

'Sharks deserve protection and I am proud to join with fellow survivors to carry that message. If we see the value in saving these animals after what we have endured, then everyone should,' said Debbie Salamone.

CHAPTER 11

WHERE SHARKS LIVE AND HOW TO AVOID THEM

TOP AMERICAN SHARK-ATTACK BEACHES

NEW SMYRNA BEACH, FLORIDA

New Smyrna is the shark-attack capital of the world according to the International Shark Attack File, which cites 210 attacks in the beach's home county of Volusia, Florida. There are more incidents per square mile of New Smyrna than any other beach in the world, earning it the location 'Shark Attack Capital of the World'. It is home to Tiger, Blacktip and Spinner sharks. Most of the attacks are described as 'fairly mild' or 'minor bites' (apparently sharks accidentally bump into humans and instinctively take a bite) and victims quite often drive themselves to hospital. High numbers of people in the water are leading to increased attacks; also the fact a swimmer looks very much like a turtle to a hungry shark! But none of this puts people off. Indeed, the deputy beach chief says that whenever he closes the beaches after shark sightings or attacks, he receives angry voicemail messages.

NORTH SHORE, OAHU, HAWAII

Second on the International Shark Attack File for unprovoked attacks is Oahu, where Tiger, Galapagos and Sandbar sharks all congregate in high numbers, especially near beaches on the island's north shore.

LONG BEACH ISLAND, NEW JERSEY

The idea for the book and 1974 film *Jaws* came from incidents at this New Jersey beach in 1916 when, in just11 days, five major shark attacks took place along the Jersey Shore, four of them fatal. Reports cited blood turning the water red and sharks following victims towards the beach. Today, sharks are rarely seen here but the legend lives on.

WEST END, GRAND BAHAMAS ISLAND

The death of Austrian lawyer Markus Groh, who was diving with sharks off the Bahamas in February 2008, put the focus on these tropical waters. One diver described the area as 'among the shark-iest places on the planet' after seeing 4.2m (14ft) Tiger sharks offshore from beaches where millions of tourists swim each year. These waters are also home to Hammerheads, Bull sharks and Blacktips. The area has become famous for shark cage diving, drawing many tourists who actually want to come face to face with a hungry shark (protected by a metal cage, of course – albeit protection that Groh did not have when he was attacked).

STINSON BEACH, CALIFORNIA

In the shadow of Marine County's Mt Tamalpais, Stinson Beach is a spot where Great White sharks swim into the shallows to feed on seals – and intimidate surfers. One tour operator described the area as 'the granddaddy of all shark beaches.' Since 1926, there have been 96 shark attacks, with seven fatalities in the state of California.

BEACHES OF BREVARD COUNTY, FLORIDA

There have been 90 reported shark encounters and one fatality here over the last 100 years. Of the 71 that happened worldwide in 2007, 32 of them occurred in Florida. It is said that the state has so many shark attacks 'simply because it has a lot of tourists' and Brevard County is just an hour's drive away from Disney World. The seas of Brevard County are also easy pickings for sharks as this is one of the most dangerous places for drowning in the rip current.

HORRY COUNTY, SOUTH CAROLINA

Over the past century, South Carolina has seen more than 50 total shark attacks according to the International Shark Attack File. Of those, 16 attacks are recorded off the beaches of Horry County, where the town of Myrtle Beach is famous as a tourist destination. However, there have been no fatal attacks in the county since 1852. Almost 40 species of shark are indigenous to South Carolina. The species are generally mild, including the Sandbar and Bonnethead sharks, but more aggressive species, including the Tiger and the Bull shark have been spotted. South Carolina's offshore estuaries provide good birthing and feeding grounds for these sharks.

SOLANA BEACH, CALIFORNIA

Solana Beach, home to a population of seals, is at the periphery of the corridor where great sharks commonly roam. A freak Great White attack in 2008 at Solana Beach in San

Diego County, California killed a 66-year-old swimmer. He was on a morning swim, training with a group, when the attack occurred.

BOLINAS, CALIFORNIA

A small beach enclave located in Marin County in northern California, Bolinas is known for its reclusive residents (cooking guru Martha Stewart lives there and the place even has its own exclusive dating site), a feel-good seaside vibe and, well, Great White sharks. In an area known as the 'Red Triangle' in Northern California (which includes Bolinas and nearby Stinson Beach), Great Whites are known to patrol the waters in large numbers. In 2002, a 3.6–4.2m (12–14ft) Great White shark jumped out from the water and snapped a screaming, 24-year-old surfer in its jaws. The man, who miraculously survived the attack, nevertheless needed 100 stitches to close his four bite wounds. He was certainly not the first (nor the last) surfer to be attacked along this stretch of Californian coast. Surfers here are already tough from braving the cold waters, but doubly courageous for braving the sharks. Plus, the wetsuits necessary in Bolinas make surfers look a whole lot like a Great White's favourite meal: sea lion.

GALVESTON, TEXAS

The International Shark Attack File cites one fatality and 12 attacks since 1911 at the beaches of Galveston.

TOP SHARK-ATTACK BEACHES IN THE REST OF THE WORLD

ZIHUATANEJO, MEXICO

Over the course of a month in the spring of 2003, the beaches near the city of Zihuatanejo, on the Pacific Coast north of Acapulco, saw three shark attacks and two fatalities. A shark hunt ensued.

GANSBAAI, SOUTH AFRICA ('SHARK ALLEY')

A popular holiday resort and fishing town, Gansbaai is another of the world's Great White shark haunts. In fact, there are so many of them that the narrow sea channel between Geyser Rock and Dyer Island is known as 'Shark Alley'. Shark cage diving has become a popular tourist attraction here. Despite the danger, many visitors go to Gansbaai just to catch sight of a Great White.

KOSI BAY, KWAZULU-NATAL, SOUTH AFRICA

Kosi Bay is a picturesque and dramatic series of four lakes in the preserved KwaZulu-Natal, which connect to the Indian Ocean. Yes, it is beautiful – but it is also famous for its Zambezi sharks (or Bull sharks, as they are more commonly named), which are known to swim into freshwater lakes and estuaries in search of food. Zambezi are not only the most aggressive of all sharks, they have the ability to swim deep inland by way of lakes and rivers. In fact, not only are these sharks found aplenty in Kosi Bay, they have been spotted swimming as far from the ocean as Ohio, up the Mississippi River in America. Sharks in lakes are a particularly scary concept.

RECIFE, BRAZIL

Sometimes called the 'Venice of South America', this seaside town on the Atlantic coast of Brazil has the perfect urban beach, with gentle seashore breezes, endless days of sunshine and... sharks, lots of them. Since 1992, there have been more than 50 shark attacks –16 of them fatal – along a 20km (12-mile) stretch of coast near Recife. This makes Recife the world's most fatal place for shark attacks, where around one in every three ends in death. Because of the concentration of the aggressive and dangerous Bull shark in Brazil, this beach has earned a reputation as being one of the most deadly shark-attack spots in the world.

LAKE NICARAGUA, NICARAGUA

Some say the largest lake in Central America does not deserve its reputation for being a shark hot spot but there was a time when literally thousands of Caribbean Bull sharks lived here. The shark's high tolerance of fresh water enabled this predator to adapt to the waters of the San Juan River, which meant it could travel up the river and then reach the lake, terrifying fishermen and other locals.

BONDI BEACH, NEW SOUTH WALES, AUSTRALIA

The presence of protective nets is a good indication that there is a serious threat from sharks here. These have been put in place to keep sharks away from humans enjoying watery pursuits close to the shore. Recent attacks have pushed the beach high up on the list for being one of the world's most shark-infested beaches. In 2006, a young woman was attacked and killed by three Bull sharks just north of this beach. Then, in 2008, a 16-year-old bodyboarder was mauled and killed by a Bull shark. Months later, a surfer and two divers had close run-ins with sharks but thankfully, lived to tell the tale. With lots of sharks and also lots of people in the water, the eastern coastline of Australia has some of the highest shark-attack concentrations in the entire world.

Almost every avid surfer along this stretch of coast has his or her own shark story to relate, or knows someone who does.

REUNION ISLAND

Home to stunning waterfalls, lush green landscape, beautiful beaches, great surf... and lots of sharks! Although there have been just a few attacks off the coast of Reunion Island in the past few years, the number per capita of swimmers, surfers and divers on this tiny, isolated island is one of the highest in the world. Since 1980, there have been 24 shark attacks off the shores of this island, 13 of them fatal. Both Bull and Tiger sharks trawl the waters.

UMHLANGA ROCKS, SOUTH AFRICA

These stretches of beautiful beach are also home to lots of sharks, including the terrifying Great White and Bull sharks. Swimmers here are somewhat protected by a string of fishing nets that keep sharks out. These were installed in 1957 after five attacks in three months. The presence of so many sharks here prompted scientists in Umhlanga to patent a device to be worn by surfers and divers called a Protective Oceanic Device (POD), which supposedly prevents sharks from attacking. The device worked by enveloping the wearer in a 120-volt electrical field, repelling any nearby sharks. Unfortunately, users did not always activate the POD until after they had spotted a fin in the water – with sometimes fatal results. The town is now home to the Natal Shark Board, which comprises a museum, headquarters and laboratory studying sharks in the region.

Mystery of the 'Jumping' Sharks

For many years, the Lake Nicaragua shark population presented a mystery for scientists: they were unable to understand how the sharks ended up in the freshwater lake, assuming they must have been trapped there centuries ago. However, in the 1960s, scientists found that these ferocious sharks jumped upstream in the San Juan River, much like salmon. Over time, younger generations adapted to the fresh water until the sharks were able to reproduce in fresh water with no need to travel back to the salty Caribbean waters. Hence, the shark became a permanent inhabitant of the San Juan River and Lake Nicaragua. However, when a Japanese shark-fin processing plant was built on the shores of the river in the 1970s, thousands and thousands of sharks were caught and killed each year, causing the population to sharply decline. The last media reports of a sighting date back to the year 2000, and although inhabitants of the San Juan River banks report sightings of sharks every now and then, there have been few recent scientific investigations and the shark population is considered to be virtually wiped out.

In May 2010, The Art of Manliness website published its own 'manly' advice about shark attacks:

> There are two types of shark attack: provoked and unprovoked. Provoked attacks occur when the human touches the shark first. These usually happen when some knucklehead scuba diver tries to feed a shark or grab its

tail on a dare. If you're dumb enough to grab a shark by its tail, you deserve whatever you get. Sorry.

Unprovoked attacks happen when you're just chilling on your surfboard and a shark swims up, bites your leg, and pulls you down into the water *Jaws*-style. Why do sharks attack humans? It's probably not for food. Humans don't make a good meal for sharks because we don't have the fat that sharks need to power their huge, scary bodies. It's more likely the shark is just figuring out what you are. Unlike most animals that check things out by looking at the object or smelling it, sharks just bite the hell out of whatever they're exploring. It's messy, but it gets the job done.

In the area of unprovoked shark attacks, scientists have observed three different kinds: the hit and run, the bump and bite, and the sneak attack:

The Hit-and-Run. This is the most common type of attack. It occurs in surf zones, where swimmers and surfers are easy targets. The victim usually doesn't see the shark before he feels its teeth sinking into his flesh. After sneaking up on the victim like a ninja, the shark will take one bite and then swim off, never returning. Why do sharks do this? Because they can, that's why. But seriously, he was probably just curious about what you were and wanted to find out by taking a chunk of your leg. After sampling your bad-tasting meat, he decided to find lunch somewhere else.

Your chances of surviving a hit-and-run attack are pretty good, barring the shark hitting any vital organs with his bite, and if you get medical attention immediately.

Bump-and-Bite. Unlike the Hit-and-Run, the victim will often see the shark before it attacks. The shark will start off by circling its potential victim and giving him a few bumps with its snout. You know, to mess with you. After sufficiently scaring the crap out of you, the shark will start biting you. Repeatedly.

Bump-and-Bite attacks result in severe injuries or fatalities. They usually occur in deeper waters, but can sometimes occur near shore.

Sneak Attacks. Sneak Attacks are sort of like the Hit-and-Run – the victim usually can't see the shark before the attack. But unlike the Hit-and-Run attack, where the shark will take a gnaw of your arm and then swim off, in Sneak Attacks the shark will bite repeatedly. Usually until you die.

Sharks – Right on Target

Results of a study published in March 2011 revealed that sharks have a 'mental map' to navigate through waters. Tracking data of three species of shark provided researchers at the University of Florida with the first evidence that some of them swim directly to targeted locations. Researchers re-analysed tracking data from acoustic transmitters on nine Tiger sharks off the south shore of Oahu, Hawaii in 1999, nine Blacktip Reef sharks in the lagoons of Palmyra Atoll in the Central Pacific Ocean (2009) and 15 Thresher sharks off the southern California coast (2010). The creatures were followed for at least seven hours and the statistical analysis determined whether the sharks were moving randomly or towards a known goal.

Said marine biologist Yannis Papastamtiou: 'There's been several studies that have shown that marine predators, like sharks, penguins, turtles and tunas, move using particular types of random walk, but there's going to be times when these animals don't move randomly. This study shows that at times sharks are able to orient to specific features, and in the case of Tiger sharks, the distance over which they're performing those directed walks is likely larger than the distance of the immediate range of their sensory systems.' He added that the research would potentially be useful for obtaining accurate population dispersal models for the sharks so that movement patterns can be predicted after changes caused by fishing or the relocation of prey.

IF YOU SHOULD MEET A SHARK...

With so many shark-infested countries, it's only right that they should draw up their 'If you should meet a shark...' own shark-attack advice. The basics remain the same, so it's only wise to swot up on the advice given below.

- If you are in the water, remain calm. You cannot outrun a shark and sharks can sense fear.
- Keep your eye on the shark at all times. Sharks may retreat temporarily and then try to sneak up on you.
- If you can't get out of the water right away, try to reduce the shark's possible angles of attack.
- Fight – playing dead doesn't work. A hard blow to the shark's gills, eyes or, as a last resort, to the tip of its nose will cause the shark to retreat. If a shark

continues to attack, or if it has you in its mouth, hit these areas repeatedly with hard jabs and claw at the eyes and gills.

- If you are near shore, swim quickly, but smoothly. Thrashing will attract the shark's attention.
- Sharks have difficulty biting things that are vertical (their nose gets in the way), so avoid leaving your hands and feet loose or going horizontal to swim away from the shark.
- Sharks can't breathe out of water, so, if possible, hold the bitten part of your body out of the water and get their gills into the air and they will let go of you.
- Sharks tend to thrash prey around to tear chunks out of it, so you should latch on to the shark.
- Repress the urge to scream. Screaming will not deter the shark much and may provoke it further.

On 12 January 2010, three people were bitten while swimming off Quy Nhon, Vietnam. Local fishermen say they sometimes catch nham fish (a kind of shark), weighing around 10–15kg (22–33lb) each, and this was probably the source of the attack.

CHAPTER 12

SOUTH AFRICAN SHARK ATTACKS

'THE SEA AROUND HIM TURNED RED...'

Friends watched in horror as Tim Van Heerden, 49, was literally ripped from his surfboard and into the jaws of a shark on 23 August 2011. It was obvious that no one could survive such a frenzied attack. Indeed, Van Heerden died within moments of arriving at hospital after suffering extensive blood loss from his injuries. The shark had bitten his left leg twice, high on the thigh and in the groin, before swimming off in the waters of popular Plettenberg Bay.

Local resident Cameron Payne, 18, was with Van Heerden and four other surfers when the shark attacked. He said: 'I was about 20 metres [65ft] away from Tim; I was just lining up a wave when I heard one of the two Australian guys surfing with us shout "Shark!" I looked across and saw the shark's tail thrashing as it churned up the water around him – there was a lot of blood.' Van Heerden's close friend Tim Clarke, who was surfing with him, added: 'He was lying on the board paddling back to the surf after a ride when I heard him scream as the shark hit him the first time and pulled him off the board. Tim was trying to climb back on when the shark came around and hit him again. I only saw the fin. Tim disappeared under the water for a moment and when he came up a few seconds later, the sea around him turned red.' Charlie Reitz, another friend of Van Heerden's, also witnessed the attack and swam out to pull the severely injured surfer from the water as he drifted, still clinging to his surfboard, and take him to shore. Said Reitz: 'He was in really bad shape when I reached him – I think he had already bled out. He was not focusing and his eyes were glazed over.'

The National Sea Rescue Institute (NSRI) had advised bathers and surfers against entering the sea in the Plettenberg Bay area. NSRI spokesman Craig Lambinon said Van Heerden was unconscious when volunteers and emergency service personnel arrived at the scene, the victim's heartbeat failed and was restored a number of times as rescue workers struggled to keep him alive. Lambinon added: 'Extensive resuscitation efforts commenced at the scene, in the ambulance and at the hospital, but he was declared dead by doctors.'

None of the witnesses were able to positively identify the type of shark that had attacked Van Heerden, but there were suggestions it could have been a Great White as these sharks are found in the Robberg Peninsula, attracted by a large seal colony. Van Heerden, known locally as 'Tim Boots' because of his job as a shoe maker, left behind a son, Jethro (17), and his wife, Michelle Marx, who said: 'Surfing was his life – that's how he wanted to live and that's how he wanted to die.'

- On 21 May 2011, spearfisherman Warren Smart, 28, died after being bitten by a shark at Cape Vidal in northern KwaZulu-Natal. While being treated for severe blood loss from a leg injury, Smart said he thought the shark had intended to grab the fish he had just shot. He died shortly afterwards. His friend Trevor Hutton commented: 'Warren was removing the fish from his spear when the shark grabbed his thigh instead of the fish, which it wanted. He wasn't the target – Warren even said this to us. He said it was an accident and that the shark may not have meant to attack him.'

- Zama Ndamase, 16, was fatally attacked by an unidentified shark on 15 January 2011 as he surfed with his brother Avuyile and other members of the local surf club at Second Beach, Port St Johns, on the Eastern Cape. Witnesses said Ndamase managed to catch a wave after being bitten and attempted to reach the shore but he bled to death in the water before being reached by lifeguards and rescue craft. The young lad was one of the most talented surfers to emerge from that remote region and had been awarded an SSA Surfing Scholarship in 2009.

  Malcome Logie of the Border Surfriders Association (BSA) said: 'Zama was a young guy, full of spirit and always ready for a laugh. He was always willing to help his teammates and enjoyed giving the younger surfers encouragement and advice. Border was looking to him to play a leading role in our team this year. His passing leaves us numb and with a huge sense of loss.' Online newspaper *ZigZag News* said that there was concern that the local community would be unable to cope with the tragedy and a counsellor was being sent in to support them. The report added: 'The Port St Johns community is poor and there is precious little money to assist the Ndamase family deal with their grief. Nor are there funds available to pay for the costs of sending a counsellor to help the community. The Border Surfriders Association and Surfing South Africa are therefore appealing for support to cover these expenses. All funds raised will be used to assist the Ndamase family with expenses and contribute to the costs of sending a grief counsellor to work with the family and the community.'

- On 21 September 2010, Khanyisile Momoza, 29, paid dearly for poaching perlemoen – also known as abalone and a valuable shellfish prized for its fleshy insides akin to oysters. He was fatally attacked by a Great White in the waters near Gansbaii on the Western Cape. Momoza was one of 12 poachers who had tried to swim to safety after spotting the shark in the shallow waters. A friend who witnessed the attack recounted: 'There was screaming and crying. We just swam, we didn't look back. We were swimming in a group but he was a bit behind us. It jumped out of the water with him and then it took him down.'

Illegal harvesting of perlemoen is big business in South Africa, where the valuable shellfish are common along coastal areas, but widespread farming of the shells has sparked fears the population could decrease. In 2007, South African authorities listed the species as endangered with the global wildlife protection body CITES. The tight restrictions were lifted in July 2010, although it remains illegal to harvest perlemoen without a licence. Many poor workers risk arrest or injury to hunt for the wild shells, whose meat can be worth up to £25 a kilo ($39). In 2008, there were reports that a group of four poachers had been taken by sharks off Gansbaii.

- On 12 January 2010, Lloyd Skinner, 37, was literally eaten by a Great White as he stood in chest-deep water off Fish Hoek Beach, Cape Town. His diving goggles and a dark patch of blood were all that remained in the water. Hours earlier, Cape Town's disaster management services had issued a warning that sharks had been spotted in the water, but the shark flag was not flying.

The Great White went for Skinner not once, but twice. Witness Gregg Coppen described the horror: 'Holy shit! We just saw a gigantic shark eat what looked like a person in front of our house. That shark was huge, like dinosaur huge. There was this giant shadow heading to something colourful, then it sort of came out the water and took this colourful lump and went off with it. You could see its whole jaw wrap around the thing, which turned out to be a person.' British visitor Phyllis McCartain added: 'We saw the shark come back twice. It had the man's body in its mouth and his arm was in the air, then the sea was full of blood.'

Kathy Geldenhuys, who had been sitting on a nearby bench, said: 'My husband had just pointed out how far the man was swimming from the other people. He asked what would happen if he was attacked by a shark because he was so far away. The words were hardly cold when the shark attacked that man. The shark attacked twice; it turned and attacked the man again. I just saw the blood on the water.'

Four rescue boats and a helicopter searched in vain for Skinner but Ian Klopper, a spokesman for the National Sea Rescue Institute, said: 'You can rule out any chance of finding him alive. Whether we find body parts, it's very unlikely – we think the shark took everything.'

According to shark scientist Alison Kock: 'More than 70 per cent of recorded Great White attacks on humans result in just the shark biting and then leaving. There is that 30 per cent where the shark behaves like it did in this case, where it came back and killed the person.' Kock added there had been an increase in shark sightings in the weeks leading up to the fatal attack.

- Lifeguard Tshintshekile Nduva, 22, was pulled off his paddle board by a shark and died almost instantly. The attack happened on 18 December 2009 at Second Beach, Port St Johns on the Wild Coast. Fellow lifeguard Abongile Maza, who was paddling a few metres behind Nduva, said the attack had been sudden and vicious, adding the water had turned blood red. 'I could not believe what I saw with my own eyes because it did not seem real,' he said.

  Police said none of Nduva's remains had been

found. Second Beach remained closed for the weekend after the attack. The incident provoked a call for more protection from sharks. Khaya Mjo, Wild Coast Guards managing director, asked: 'Why are only lifeguards the ones being attacked while nothing happens to ordinary people? Why do these attacks happen at the same beach?'

- On 29 August 2009, Gerhard van Zyl, 25, died in hospital following a shark attack at Glentana Bay in the Western Cape Province. He had been surfing with a friend at the bay between Glentana and Outeniqua, near Mossel Bay. National Sea Rescue Institute spokesperson Craig Lambinon said they were called out around 3.30pm, soon after the attack, adding: 'We found him on the beach and treated him for amputation.' Van Zyl's lower right leg had been bitten almost completely off. 'The victim was airlifted to a hospital in George in a critical condition, where he was later declared dead after all efforts to save him were exhausted,' added Lambinon.

  Robin de Kock, general manager of Surfing South Africa, commented: 'This is a terrible tragedy and without detracting from that fact, it must be said that it is a very rare occurrence. If you're a surfer, shark attacks are one of the risks you're aware of. You could be stung by a jellyfish or break your neck on a wave, but surfing is still very safe compared to contact sports like rugby.'
- Luyolo Mangele, 16, died after being bitten on the leg by a shark while surfing off Second Beach, Port St Johns on 21 March 2009. His attacker was believed

to have been a Zambezi (Bull) shark. The other two surfers in the water at the time said they heard screams but could only find a pool of red water. Mangele made it to shore but then lost consciousness. Concerns were raised that heightened shark activity in the area could be due to the influx of sewage-loaded water from a river. Following the attack, authorities increased the number of lifesavers and coastal patrollers to watch for sharks.

- On 24 January 2009, off-duty lifeguard Sikhanyiso Bangilizwe died an horrific death in the jaws of a shark while swimming off Second Beach, Port St Johns on South Africa's Wild Coast. It was believed a Bull shark was behind the attack, which ripped Bangilizwe's body in half, taking away his right arm, shoulder, back and buttock. One moment he had been catching the waves, the next he was literally torn in half by the frenzied beast. The horror was witnessed by a friend, fellow lifeguard Tshintshekile Nduve: 'He was catching waves in the water further from where I was, and when we passed the waves, I heard his cries. I saw he was in trouble and the shark on him. I saw blood and I went out of the water to get help.'

Bangilizwe, 25, died from massive injuries: his body was bitten in two and he lost his right arm, shoulder, back and buttock. Vuyo Maza, one of the lifeguards who sped to the scene by boat, said he was at a tower when the attack took place: 'When we went out to the water, Bangilizwe was dead. His body was badly injured and we could see his insides

– it was the scariest thing I have ever seen. To see him die has made it difficult for me to sleep. I wish I could get pills so that I don't get troubled. I am not well, but I will continue to go into the water as a lifeguard.'

The attack was believed to have been by a Bull shark. A member of the Wild Coast Guards, Bangilizwe was the second lifeguard in two years to be killed by a shark in that location.

- On 14 January 2007, Sibulele Masiza, 24, was killed by what was believed to be a Tiger shark while surfing at Fishery Bay, Port St Johns. 'Judging by the circumstances of his disappearance and the flipper that was torn, it can only have been a Tiger shark attack,' said Jeremy Cliff of the Natal Sharks Board. Masiza's torn flipper was washed up onto the beach soon after he disappeared, but his body was never retrieved. Cliff said Port St Johns was a popular spot for the small fish that sharks feed on. Before the fatal attack, Masiza was already listed in the Natal Sharks Board records: he had been bitten on the legs by a shark in 2004.
- Lorenzo Kroutz, 17, was believed to have been taken by a shark (and subsequently eaten) on 22 March 2006 while swimming at Port Alfred in the Eastern Cape Province but there were no witnesses.
- On 14 August 2005, a human foot was recovered from the waters at Minerton Lagoon, Cape Town.
- Diver Henri Murray, 22, was fatally attacked at Miller's Point in the Western Cape Province on 4 June 2005. A fellow diver managed to shoot at the shark

with his speargun but the shark remained undeterred. The last the diver saw of his friend was in the jaws of the shark. A search was launched in a bid to find the student's body, but only the floatation buoy (which attaches to a speargun) was discovered. Police divers also found the speargun, a flipper, a mask, a snorkel and parts of a weight belt belonging to Murray. The top of his red wetsuit later washed ashore, with the keys to his car still in one of the pockets.

- On 10 March 2005, Anthony Arnachallan, 51, was attacked and killed by a Tiger shark while in the waters of Isipingo, KwaZulu-Natal. His body, bearing the savagery of the attack, was later recovered. Jace Govender of the Nokia rescue helicopter said there were three bite marks on the body, adding: 'The sea is dirty and we suspect the shark was drawn to the area by the dirty brown water.'
- Grandmother Tyna Webb, 77, was literally swallowed up in the jaws of a 6.7m (22ft) Great White shark as she took a morning dip off Fish Hoek Beach, Cape Town on 15 November 2004. Witnesses, including children, saw the attack and watched in horror as the shark circled Mrs Webb three times before dragging her beneath the waves. She was only 18m (60ft) from the shore. One onlooker, Paul Dennett, observed the nightmare scene from his beachside home, along with his fiancée and her daughter. He said: 'I saw a wild thrashing in the sea and at first thought the shark was attacking a seal. Then I saw somebody in the water. There was a hell of a lot of blood. I then saw the shark circle around

the victim and then, in just one big mouth and not even breaching the water, took her in.'

Another witness said the shark, '...took her, leaving her lying in the water, and then came back for her, again and again.' It was a tragic end for the woman who had bathed at that same spot every day for 17 years. Rescuers who searched in vain said they had spotted a shark 'as big as a helicopter'. An air-sea rescue mission failed to find a body – all that was left of the incident was Mrs Webb's red swimming cap.

- On 2 June 2004, Nkosinathi Mayaba, 24, was fatally attacked while swimming at Dyer Island, Western Cape Province. His left leg was severed in the attack by a Great White.
- Seldon Jee, 21, disappeared while spearfishing at Sodwana, KwaZulu-Natal on 27 November 2003. A 4m (13ft) Tiger shark had been seen in the area. Jee's body was never recovered, although a severed hand was identified as being his.
- On 12 September 2003, David Borman, 19, died in a shark attack while he was bodyboarding off Noordhoek, Western Cape Province. He was reportedly no more than 30–46m (100–150ft) from shore when attacked by a Great White, which might have mistaken his black wetsuit-clad body and fins for a seal. Borman died from what was described as 'a massive injury from his back down to his thigh.' Grisly photographs of what appeared to be his blood-drained body circulated on the internet with some saying the pictures were fake.

- An unidentified male, aged around 14, was said to have been taken by a shark scavenging near a lagoon at Table View, Western Cape Province on 17 January 2003.

Not all of South Africa's shark attacks are fatal, of course but there have been a huge number over the last decade and many have been pretty horrific:

- On 22 July 2011, Denver Struwig, 29, suffered severe bites while surfing at Cintza Beach. The shark first bit Struwig's leg and then his shoulder and chest, pinning him down on the seabed. He was rushed to hospital to undergo treatment to puncture wounds around his back, neck, stomach and upper and lower legs, and surgery to the laceration on his right lower leg. Local surfer Murray Elliott, who witnessed the attack, said he had been wearing a shark pod: 'I saw the shark come back three or four times for Denver. It eventually dragged him under the water and I thought that was it. I am still shaken up, but I believe the Shark Shield is what saved Denver's life.'

  The Chintsa and Surfing community hosted a fundraising event to help with Struwig's medical costs.
- On 28 June 2011, Paolo Stanchi, 22, was bitten on the leg and hands by a Dusky shark as he took part in an organised dive at Aliwal Shoal on 28 June 2011. Mark Addison, owner of Blue Wilderness Dive Expeditions, believed the attack to be a case of mistaken identity: 'From the debrief it seems that the

shark went at his fins, which were spilt fins with black and grey contrasting stripes down the fin blade and that certainly sounds like that contributed to a mistaken identity bite, where the shark thought perhaps it was biting into a shoal of fish.'

- Clinton Nelson, 33, escaped injury when a Great White attacked him as he surfed at Plettenberg Bay on 29 May 2011. His board was badly chewed.
- On 4 May 2011, Trevor Burger, 37, had to have his left leg amputated when he was attacked by a shark while spearfishing at Palm Beach, KwaZulu-Natal. He also suffered severe injuries to his hands. His friend Fourie Combrinck was present at the time and recalled the attack: 'We were both eager to shoot some fish. Trevor beat me to it and soon bagged a nice bronzy. After another fruitless dive down, I surfaced to see Trevor sitting on the boat. He shouted to me to get to the boat as there was a shark in the vicinity. This was strange to me as we have both encountered sharks before without them hassling us. I was surprised to hear Trevor say that he was bitten by the shark. According to him, at about 10 metres [33ft] and on his way to the surface, he felt the shark biting on his lower left leg and shaking him twice before letting go and swimming away. He only saw the shark after it let go of him from behind and was unable to recognise the species, but we both suspect it to be a Dusky as we have encountered them in the area before. Again, I was surprised by the speed of these animals underwater how quickly it attacked and disappeared.'
- On 4 July 2010, Sarah Haiden, 21, was left with

severe cuts to her left leg when a shark attacked as she snorkelled in the waters at Sodwana Bay in northern Zululand. She said: 'While the guys were diving, a friend and I were snorkelling. My friend got back into the boat and I was swimming to the buoy, where the divers had gone down. I started swimming back to the boat when I felt something bump me. At first I thought one of the divers was playing a joke on me but realised that they were too deep to have come up so quickly. The next thing I saw this shark in front of me as it went for my legs. I kicked out at it as I had heard that you should try and kick them in the face to chase them away. It grabbed my left leg, and I kept kicking and screaming like a banshee. Luckily, for some reason it let go of my leg and swam away.'

- Brendan Denton, 33, had both feet bitten while surfing near East Beach, Port Alfred on 13 April 2010. He said at first he thought a friend was fooling around but then saw his foot in the shark's mouth.
- On 16 February 2010, Michall du Plessis suffered lacerations to his right leg while surfing at Yellow Sands Point. Logan Philpott, who was one of the group surfing with him, said they had been in the water for an hour and a half when he heard his friend shout 'Shark! Shark!' He added: 'I thought he was joking at first, but then everyone else started shouting and when I looked, I saw this thing, a tail and a fin, shaking itself against my friend.'
- Simon Bruce, 20, was bitten on the foot while wading in water at Port Alfred on 29 December 2009. Bruce said of the attack: 'The water was a little bit murky,

but it was really flat. I was walking on a sandbank and I was no more than chest deep when I felt a bump on my foot; I didn't feel the bite at all. Then I felt the shark rub against the top of my legs and I freaked. I assumed straightaway it was a shark.'

Two friends on the beach wrapped his foot in a towel and took him to hospital.

- On 11 August 2009, Jeandre Nagel escaped injury but had his surfboard bitten by a shark in Richard's Bay, KwaZulu-Natal.
- Paul Buckley, 37, desperately clung onto an attacking shark's fin as he surfed at Jogensfontein, Stilbaai in the Western Cape Province on 7 July 2009. He said: 'I was actually paddling back to shore because I hadn't had that great surfing when I was flipped in the air with such force, I just knew in my gut that it was a shark. It was like a 500lb Rottweiler in a very bad mood; the force was incredible. The first thing I said when I was attacked was, "No, please God, not like this!" I didn't see the shark's eyes because his back was towards me but that was when I grabbed its tail. No doubt I was scared. I feared for my life so I just grabbed it. I thought if I held it by the tail, its mouth could not reach me again.'

  After a few seconds the shark released Buckley and he was able to make it back to shore. He was rushed to hospital, where he had 150 stitches to his wounds. 'If it had been a little further over to the left or right, or if it had taken out a chunk, it could have been much worse,' he added.
- On 27 March 2009, Tony Dell, 59, was bitten on the

calf while helping out at an angling competition at the mouth of the Sunday River in the Western Cape Province. Dell was measuring a Ragged Tooth shark when a wave came in and lifted it. The beast spun round and bit him.

- Gabriel Fernandez, 40, was tuna fishing off Cape Point when a Blue shark attacked his arm and hand on 2 March 2009. The shark had become caught on the line. Fernandez said: 'I was pulling the shark in, and it turned around and bit me instead.'

On 10 September 2008, Luke Parker, 15, had to have 45 stitches in a leg wound after being attacked while night-time fishing at Plettenberg Bay. He had been trying to pull in a 2m (6.5ft) Tiger shark when his line broke and it pulled him off rocks into shallow water. Ray Farnham, NSRI station commander at Plettenberg Bay, watched the whole incident: 'I stood watching the guy and thought, "The thing is going to bite you!"'

- Kobus Maritz, 46, was attacked by a Great White on 28 June 2008 while surfing at Mossel Bay Harbour. He escaped injury when the shark took a bite out of his surfboard and swam off.
- On 26 June 2008, Kevin Macghie was pulled along by a shark and had his swim fin bitten while spearfishing at Struis Bay.
- On 1 May 2008, the NSRI rescued a fisherman from a Japanese trawler after he was bitten by a shark that was caught in the net. The bite wound extended completely around the man's left leg!
- Jacques Peens, 38, lost part of his leg after he was

bitten by an unidentified shark accidentally caught while fishing at Scottburgh, KwaZulu-Natal on 28 March 2008. He said: 'I'm not really too fond of blood. Usually I would've fainted but I managed to keep my wits about me this time.' Peens had to undergo a three-hour operation on his leg.

- On 29 January 2008, Wayne Symington, 42, had his surfski bitten by a Blacktip shark at Suncoast Pirates Beach, Durban.
- On 7 November 2007, surfer Andrew Smith, 14, kicked his feet free from the jaws of a shark at an area known as The Strand in the Western Cape Province. After he had freed himself, the shark attacked his surfboard. Smith was rushed to hospital, where he was treated for lacerations to his legs and a severed tendon. His father, Jimmy, said: 'The shark took him from behind. He saw it and at one stage it seemed like it was on top of him. He said to me, "The shark was bigger than I am, and I'm six foot tall!" He said he just started kicking and fighting; his leash snapped and it tried to drag him under but he managed to get his feet out of its mouth. He was very lucky.'

  Vergelegen Medi-Clinic reconstructive plastic surgeon Dr Jonathan Toogood reported that Smith had 'hundreds of little lacerations and tooth marks on his feet and lower legs.'
- On 3 November 2007, Lee Mellin, 37, was attacked by a Great White while surfing with his friend Leigh Stolworthy at Bonza Bay. He sustained a 39cm (15in) wound to his left thigh. Describing the attack, Mellin said: 'There was nothing we could do – it just popped

up between us. There was this massive white shark bursting out of the water, real *Jaws*-like. It obviously gave one look at Leigh, thinking he didn't have enough meat so he went at me. It was a flippin' big thing! I was still on my board and it came for me – I just felt its jaws sinking in. I think the shark's teeth got stuck in the surfboard. It then took another bite, but by then I let go of my board. It just bit the board again.

'It all happened in seconds. Afterwards, we got onto our boards and paddled back. I was worried it would come back and strike again, but Leigh kept assuring me that it was nowhere in sight. The doctor said I'm the luckiest shark attack victim he has ever seen and I definitely agree.'

- In June 1007, two unidentified divers were attacked in separate incidents. One was bitten on the head by a Tiger shark while on a supervised shark dive at Aliwal Shoal and the other suffered several injuries when attacked by a Great White while scouting for crayfish in the Mtwalume area.
- On 6 December 2006, Peter Willoughby, 25, had to have 75 stitches to wounds on his hands, calf and thigh after being attacked by a Sand Tiger shark while fishing at Richards Bay, KwaZulu-Natal. Mind you, he was actually fishing for sharks at the time, even though he was trying to release this one when it fought back. Willoughby said: 'The shark could now just about swim again and I was sitting in the water when suddenly I felt something hit my leg. I felt the shark's mouth close on my leg and realised I was in trouble. Fortunately it didn't bite down as it would

have taken a big piece out of my leg, if not bitten it off. I pushed it off my leg, cutting my thumb on one of its teeth, stood up and ran out the water.

'No one knew what happened and they continued to get the shark out the water so they could release it. I stood one side in the darkness and didn't want to look at my leg – I figured it wasn't too bad as there was not much pain. While they were doing this I decided to look down and saw my calf had a big gash in it and looked a bit shredded; it looked a lot worse than it felt. I called my friend over and showed him, and suddenly the people we were helping were panicking a bit. It was quite a big shark, about 2m [6.5ft] from the nose to the base of the tail (around 150kg [330lb]) and realised I could have been a lot worse off than I was! I don't blame the shark for what happened – I see it as an accident and will continue to catch, tag and release these beautiful animals.'

- On 10 November 2006, canoeist Richard Tebbutt was attacked by the Bull shark after deciding to give fisherman a hand at the Nahoon River in the Eastern Cape Province. Afterwards, he had to have 50 stitches to an arm wound. Tebbutt was paddling on the river when he saw a fisherman on the bank battling to pull the shark in. He said: 'I just jumped out of my canoe and dived down and grabbed the fish by its tail. I got a shock of my life when I saw a big Zambezi [Bull] shark, charging towards me. It quickly turned around, grabbed my left arm and lacerated it. Within a split second, I saw my blood in the river. I just smacked the shark as hard as I could with my right arm to get it to let go of me. I was in the water holding my arm, which was bleeding profusely, and the shark started coming towards me very fast. I immediately jumped out of the water onto the rocks and it missed me. I could have been dead by now. I was shocked that I had overcome the shark – its strength was unbelievable.'

- Joseph Johnston, 36, was fortunate enough to witness only a 'threat display' by a Great White while spearfishing at Millers' Point, Western Cape Province on 30 September 2006.
- On 2 September 2006, Steven Harcourt-Wood, 37, kept his life but lost his surfboard when a Great White attacked him at Noordhoek.
- Lyle Maasdorp, 19, had his surfski bitten by a shark at Fish Hoek on 28 July 2006.
- On 9 April 2006, Stuart Duffin, 15, was bodyboarding at St Francis Bay, Eastern Cape Province when he was bitten on the leg by a Ragged Tooth shark. He decided to fight back and hit the shark on the nose but missed and caught its teeth instead, ending up with tears to his hand. Eventually he struck the shark's nose and it let him go.
- Surfer Jason Noades, 15, suffered two bites to his right calf on 8 February 2006 when he kicked back at an attacking Tiger shark at Nahoon East London, Eastern Cape Province. 'I will go back in the water in another week once the injury is properly healed,' he insisted.
- On 25 January 2006, Michael Vriese, 34, was bitten in the right arm while spearfishing at Coffee Bay on

the Wild Coast. He was pulling in a fish when the shark decided to go for his catch, mauling Vriese's arm and severing two arteries; also damaging muscles and nerves on the wrist and forearm. Said Vriese: 'By the time I had swum back to the beach, I was getting pretty weak and my friends had to support me by both shoulders to get me to a car. There was blood everywhere. I don't know how much I lost, but I'm told that I got a transfusion of at least four units.'

He was treated at Mthatha hospital before being airlifted to Durban, where he underwent further surgery to repair the damaged arteries and tendons: 'I could have ended up dead. I'm lucky that my friend and I both know a bit about first aid and I was able to stop some of the blood loss myself by tying a tourniquet around my arm and loosening it every few minutes to allow some blood to get through.'

- On 1 January 2006, diver John Williams, 49, was attacked at Soetwater and suffered wounds to his fingers.
- Ashley Milford, 25, escaped with just a bitten finger while surfing at Nahoon East London on 25 November 2005.
- On 15 November 2005, policeman Ivan Gerger, 32, did not expect a Sand Tiger shark to fight back when shark fishing at Aston Bay, Eastern Cape Province. He was catching sharks for the local aquarium who export sharks across the globe (local fishermen are contracted to catch them). After biting Gerger, the shark swam off. The victim was admitted to hospital. Police spokesperson Inspector Marianette Olivier said

the attack happened as Gerger tried to pull the shark out of the water: 'It bit both his hands and his right leg. The shark bit his palm and fingers, but didn't bite anything off.'

- Trevor Wright, 52, was attacked by a Great White while kayaking at Sunny Cove, Fish Hoek on 1 October 2005. He escaped injury but his kayak was bitten. Said Wright: 'I still have this vision of the shark's open jaw and eye. I just saw the front of the boat in its mouth. I thought, "It's you or me and it's not going to be me!" If I had fallen into the water, it would have been far worse.'
- On 25 May 2005, Jay Catarall, 32, suffered severe bites to his buttocks and legs when a Sand Tiger shark pounced as he was surfing at the mouth of the Kei River. Catarall said: 'I felt something pulling me backwards – I thought it was just the current. But after the second pull, I realised it was something in the water. It wasn't a violent attack – I did not feel any of the biting, it was just a sensation. I remember repeatedly beating the shark on the nose.' He needed 100 stitches to his wounds.
- British schoolteacher Chris Sullivan, 32, was attacked while surfing at Noordhoek on 28 March 2005. Despite punching and kicking, he still had his leg sliced by the Great White. Describing the attack, Sullivan said: 'It felt like it came up slow and I saw its eyes and it looked really dark grey. It turned and I saw the underneath of its belly, then I saw its mouth. Then it grabbed hold of my leg. I started lashing out, hitting it – I think I kicked it. It's probably in a bad

way now! I pulled the leg out. It felt like a knife through butter and I thought "Oops!"'

He managed to stay on his surfboard and catch a small wave that took him back to the shore, where a local vet – who had also been surfing – applied an emergency tourniquet to his leg. Sullivan's wound required 200 stitches. Clive Mortimer, the local station commander of the National Sea Rescue Institute, said the victim was 'extremely lucky' to have escaped alive. Meanwhile, Sullivan harboured no bitterness towards his attacker, saying: 'I haven't got a problem with the shark – I was in its water and I was stupid enough to go surfing where there was a lot of sharks. I don't think it meant to eat me – I think it just fancied a nibble.' But he would later admit to having nightmares after the attack: 'In my dreams I wasn't winning the battle, it was biting me in two. Every time it came at me, I hit it. It was a grim night and I woke up with lots of sweats but now it's fading and every time it comes at me now, I don't have to hit it anymore. It looks at me with some respect and swims away.'

- On 27 November 2004, Llewellyn Maske, 20, received three bites to the foot by a Sand Tiger shark while surfing at Nahoon East London.
- Arno de Bruyn, 16, was attacked by a Sand Tiger shark while wading in shallow waters at Gonubie, Eastern Cape Province on 26 November 2004. His lower leg and foot received serious injuries.
- On 30 October 2004, Andre Hartman, 52, a shark expert and cage-dive operator, was bitten on the foot

by a baby Great White near Gansbaai, 190km (118 miles) southeast of Cape Town. Hartman had been 'chumming' – trying to attract sharks with fish bait. The team were taking clients from the Czech and Slovak Republics for a dive with the sharks. Said Hartman: 'While I was talking to a tourist from the Czech Republic, I was dangling my right foot over the side of the boat and was bitten by a baby White shark. They were just taking nice pictures and no one was watching. My foot was dangling above the water when the shark arrived. I turned my head for a few seconds when the shark took a bit at my foot (they don't really see the bait and go for the smell and my foot was there). It was my own fault, but I suffered only three puncture holes. If you play with blades you are likely to get cut, but it's only a scratch and it happens all the time. I've been cut on my finger a few times and this is the same.'

Known locally as 'Sharkman', Hartman has both his critics and followers of cage diving, with some saying the practice encourages sharks to gravitate towards humans for food and others praising the opportunity for close encounters.

- Champion surfer Wayne Monk, 34, was attacked by what was thought to be a Ragged Tooth Shark at Supertubes, Jeffrey's Bay on 9 October 2004. Monk had just hit a wave when a shark hit him: 'It grabbed my foot and pulled me down. When I came back up and looked down, I saw this big brown thing underneath me. I'm lucky – if it had been a Great White, I wouldn't have a foot.' His lucky break arrived when another wave came in and washed him onto nearby rocks. He made his own way to hospital, where he was treated for wounds to his foot.

- On 5 April 2005, John-Paul Andrew, 16, lost a leg in a Great White shark attack while surfing with a group of friends at Muizenberg. Paramedics worked on his heart for 20 minutes after he went into cardiac failure. Andrew was rushed to the Constantiaberg Medi-Clinic Hospital. Spokesperson Gail Ross said: 'His body has been through enormous trauma, so we will have to wait and see how he reacts but he is young and strong and we are very positive.' (Andrew was later taken off the critical list.)

  Vaughan Seconds of the National Sea Rescue Institute described how lifeguards rushed to help the shark victim: 'They noticed that something was wrong when they saw half of the boy's board being tossed into the air. They pulled him out of the water and when I got to him, I saw that his right leg had been completely bitten off and that he had bite marks on his remaining leg. By then he had already lost a lot of blood.' Andrew's leg was found four days after the attack: it was still attached to his surfboard leash.
- On 7 January 2004, Alan Horsfield, 22, had his foot bitten by a Sand Tiger shark while surfing at West Beach, Port Alfred.
- A Mako shark attacked a kayak and its two occupants at Karridene on 31 December 2003. Thankfully, they escaped injury.
- On 8 August 2003, Joseph Krone, 16, had his wetsuit bitten into as he surfed at Jeffrey's Bay, Eastern Cape.

His board was shredded into three pieces. Incredibly, the day after the attack, Krone came second in a surfing competition held at the very spot where he had been attacked!

- Craig Bovim, 35, was mauled by a Ragged Tooth shark off Scarborough Beach – he had been surfing and diving for crayfish on 24 December 2002. Bovim said he had seen the shark approaching him and tried to swim slowly away but then a wave came and filled his snorkel with water. That was when the shark attacked. He tried to push the shark away and at one point his arm went down the creature's throat.

  Bovim managed to swim back to the beach and was airlifted to hospital, where he underwent a four-hour operation to arm wounds. Later he went on to set up the Shark Concern Group, which lobbies for an end to the craze of cage diving – a practice which critics say encourages sharks to attack humans. Said Bovim: 'We don't know enough about the risks. Until we do, we should stop it.'
- On 3 June 2002, snorkeller Tony White, 50, had his upper arm bitten by a Copper shark off Mkhambati, between Port Edward and Port St Johns.
- Adrian Sheik, 16, had to have his leg amputated after a Bull shark attacked as he was fishing in Durban Harbour on 4 January 2002. He tried to beat off the Zambezi (Bull) shark by kicking it. The brave teenager then stabbed the shark with his fishing rod before finally escaping and swimming towards the harbour's edge. He was rushed to Durban's Addington Hospital, where his leg was amputated.

Recounting the incident, Sheik's best friend Kevin Moonsamy, who was with him at the time, said the attack itself was quick, but helping his friend to swim back to shore took about half an hour. A local fisherman assisted the boys. It was the first reported shark attack inside the harbour since the Sharks Board began keeping records of attacks, back in the 1940s.

Sheik's mother Anita said: 'I thank God my son is alive. He loves fishing and fishes almost every day,' but she added that Sheik had been reluctant to go out on the day of the attack because he felt ill and it was Moonsamy who had persuaded him. 'Maybe it was a sign, a sort of premonition,' she added.

- On 1 January 2002, Dr. Michael van Niekerk, 26, set off to surfski from the mouth of the Mlalazi River to Mtunzini beach with two other doctor friends. Stopping to wait for one of them, he dropped his legs over the side of kayak and then the shark attacked. Van Niekerk said: 'I felt something hit my right leg with a big splash. It tugged my foot, but I managed to brace myself with my paddle. I pulled my foot up and saw a gash on the side.' He and his friends said the attack served as a reminder to surfskiers not to paddle alone or at dusk – and not to dangle their legs over the sides of boats out at sea.
- David van Staden, 28, had his leg bitten by a Great White while surfing at Nahoon East London on 8 May 2001. He said: 'I first thought it was a friend just fooling around, but suddenly I saw this thing swimming around me and I paddled as fast as I could

to the shore.' Marine services chief Willie Maritz said investigation and damage to van Staden's surfboard indicated he had been attacked by a 3m (9ft 8in) Great White. 'The shark just pushed him to check whether he was something to eat,' said Maritz.

- On 8 April 2001, Dunstan Hogan, 46, was surfing at Cape St. Francis when a Great White attacked his buttock, hip and thigh. He recalled: 'I just saw this grey mass and thrashing tail fin. I didn't see the shark coming as it attacked from underneath. I suddenly felt this enormous pressure, like being gripped in a vice. It wrapped its teeth around the board and my hip, and lifted me out of the water. I was still holding onto the board and then I felt myself going under, and I was forced to the bottom and my feet hit the sand. I opened my eyes and there was a lot of white water and sand – and this big, dark shadow. I was trying to hold the board as protection, but couldn't hold onto it underwater and let go. I popped to the surface, pulled the board towards me by the ankle leash and got on it. I didn't feel any pain, probably because the water was cold, about 15 degrees.

  I started paddling to the shore and then the shark came for me again, coming from the shore – I just saw this grey mass and thrashing tail fin. When it attacked, I was lifted out of the water again. Then it left, and I started paddling again and caught a broken wave into shore. My wetsuit was all severed. I held it together and walked up on to the grass and lay there while someone went to call the doctor. Only then did I start to feel the pain.'

In fact, the shark bites had missed Hogan's main artery by just 2cm!

- On 16 July 2000, Shannon Ainslie suffered hand injuries after being attacked by two Great Whites while surfing at Nahoon East London. Recalling the attack, which was broadcast on TV channels such as National Geographic and Animal Planet, Ainslie thought he was 'in a daydream.' He said: 'I was paddling for a wave when one Great White came up on my left and one on my right. The one on the left bumped me into the air, bit my surfboard and my hand and took me underwater, while the one on my right was going for my head and shoulders, but missed me because the other shark bumped me out of the way. Every other surfer that was out at the Reef left me in the waves and I prayed to God to help me out of the water. Straightaway a wave came out of nowhere – it made me take God and my life much more seriously. It was only later that I realised two of my fingers were hanging by their skin, but they were sewn back on.'

Shocked and bleeding, he was unaware of the second shark at the time so watched in amazement, the footage later shown to him by a Canadian tourist. Said Ainslie: 'It was pretty cool, actually. It was cool because I was alive when I should have died. It was even more of a miracle that I survived an attack by two Great Whites.'

Later, he went on to compete in major South African surfing competitions.

There have been 214 shark attacks, 42 of them fatal, in South African waters in the last 100 years. Tourists still flock to dive with the sharks. This 'pastime' is blamed by some on increased shark attacks as it encourages the creatures to come closer to shore.

South Africa was one of the first countries to officially protect the Great White hence the species has increased in number. Dyer Island near Capetown has the nickname 'Shark Alley' because of its high Great White population. They particularly like hanging out at Geyser Rock, home to more than 50,000 seals.

SHARKS OF SOUTH AFRICA

- Tiger shark
- Hammerhead
- Bull shark
- Blacktip
- Great White
- Zambezi

One of the youngest shark fatalities was an unidentified boy aged seven or eight, who died while playing on rocks near the yacht club in Jakarta Harbour, Indonesia. He was caught by a Tiger shark. Also in Indonesia, a Tiger shark caught on 16 June 2002 in the local waters and offloaded at Samut Prakan, 32km (20 miles) south of Bangkok, Thailand, had a right forearm and leg in its gut. Forensic examination suggested the remains had been there for one or two weeks.

CHAPTER 13

BRUTISH SHARKS OF BRAZIL

'THE SHARK PULLED ME UNDER WITH SO MUCH POWER THAT I REALLY THOUGHT I WAS GOING TO DIE...'

Out of all South America's 105 confirmed shark attacks, 90 – with 21 fatalities – have occurred in Brazil's waters. This large number is blamed on the 1984 development of Porto Suape, to the south of Recife, an industrial port which handles more than four million tons of cargo a year. The building work sealed off two freshwater estuaries which had discharged into the Atlantic Ocean and which had served as birthing waters for many Bull sharks. The sharks headed for the next estuary instead, which discharges right into Recife's waters, and a nearby channel used by surfers became the sharks' new feeding ground. Fabio Hazin, a marine biologist and head of the state-funded monitoring committee, explained: 'Female bull sharks used to enter those estuaries to give birth. From when the port was built, we believed a number of females moved north to the next estuary, which discharges on to the stretch of beach where the attacks happened.'

Based on that finding, local human rights lawyers started to consider whether the state of Pernambuco should give compensation to shark attack victims. The Safe Beach Project also proposed a solution to Recife's shark problem – a call to the local authorities to partition off a short section of beach using heavy-duty nets out at sea. Electromagnetic buoys would also deter – but not kill – the sharks. These are similar to the measures in place in South Africa and Australia, but the idea has met with opposition in Recife.

Completing the shark paradise is a slaughterhouse which disposes of blood into the estuaries. Thanks to all of this, Recife's 12.5-mile (20-kilometre) coastline has become an extremely dangerous place, with a higher proportion of shark attacks resulting in death. One in three shark attacks that occur in Recife are fatal.

- On 1 August 2011, the body of Gabriel Alves dos Santos, 14, was recovered by lifeguards. He was last seen swimming at Pina Beach the day before and his disappearance was recorded as drowning but after the recovery of the body, a shark attack was blamed. Dos Santos' arms and legs had been bitten off!
- Maurício da Silva Monteiro, 34, was literally torn to shreds while swimming off Piedade, Recife on 13 September 2009. The next day his body was found bearing all the signs of a shark attack with one leg partially torn off and injuries to both arms, face, buttocks, back and abdomen.
- The body of an unidentified male was recovered at Goiana, Pernambuco on 10 July 2006. His killer could have been either a Bull or Tiger shark.
- On 18 June 2006, surfer Humberto Pessoa Batista, 27, died in Olinda, close to Recife, Brazil, after being

attacked by a shark at the beach of Punta del Chifre. He was with around 30 other surfers and about 15m (49ft) from the beach at the time. The young man was helped by a friend and taken to the hospital in Recife by lifeguards. But the shark bite ruptured the femoral artery in his groin and he died of a haemorrhage. Following this, the area was cleared of surfers. According to the lifeguards, it was the first shark-attack case at Punta del Chifre beach but Batista's death was the nineteenth in the general area since 1992. The sharks are attracted to a large coral reef in the waters off Recife, where they go to feed.

- Orlando Oscar da Silva, 22, was fatally attacked while swimming off Piedade Beach on 1 May 2004.
- On 29 February 2004, Edimilson Henrique dos Santos, 29, was savaged to death off Piedade Beach. He had been just 10m (33ft) away from the shore when the shark attacked his legs and hips. Dos Santos died in hospital shortly after the attack.
- Surfer Aylson Gadelha, 19, died of injuries from a shark attack after surfing off Boa Viagem Beach, Recife on 1 December 2002.
- On 14 October 2002, Luiz Soares de Arruda (36) was fatally savaged by either a Bull or Tiger shark while swimming off Piedade Beach, Jaboatao dos Guararapes City, Recife. His body was never recovered.
- Fábio Fernandes Silva, 16, was fatally attacked while swimming off Boa Viagem Beach, Recife on 23 March 2002.
- On 2 July 2001, a shark killed a young Brazilian man off a beach in the northeastern city of Recife, known

for its shark attacks. Two days after he went swimming, the body of 20-year-old student, Carlo Alberto Brasileiro, washed up on the famous Boa Viagem beach. The front of his thorax and all his internal organs were ripped out. His body was also missing a forearm, part of the right thigh and the face.

It was not clear whether Brasileiro had been swimming beyond the coral reefs that run along the coast of the city (the Pernambuco State Government warns bathers not to go beyond these reefs and in 1999, banned surfing in the area after a young man lost his two hands in a shark attack, also off Boa Viagem). This was the 33rd recorded shark attack on Pernambuco's southern coast and the eleventh fatality since 1992.

As with other shark locations, in Brazil there have been plenty of severe, but non-fatal attacks, too:

- On 2 June 2008, a shark ripped off the right hand of Wellington dos Santos, 14, when he dared to venture beyond a coral reef keeping sharks away from Piedade Beach, Recife. The teenager also had a massive chunk taken out of his buttocks and horrific pictures appeared in the local press. Fortunately, lifeguards managed to rescue dos Santos and rushed him to a hospital. Pernambuco state fire department spokesman Marcio Maia commented: 'People insist on ignoring the signposts warning of the danger of shark attacks, especially beyond the coral reefs about 490 feet from the beach,' adding that sharks have killed 19 people in Pernambuco state over the past 15 years.

- Malvis Cristino de Souza, 28, was bitten on the foot while removing a shark from a trap at Itamaracá, Pernambuco on 7 December 2007. He said: 'I pushed the slang [a piece of wood that closes the trap] with my foot and then just felt something snapping my foot. I started screaming and my friend helped to scare the animal – I was very afraid. The shark was so big, it broke through the trap.' At the hospital where de Souza was treated, Dr. Claudio Souza said: 'The attack was characteristic of being a warning when the animal wants to scare. He had no nerves hit. If it was a concentrated attack, for sure, it would have plucked the leg off the boy. Would you try to catch a shark because it makes money?'
- On 21 May 2006, Antonio Carvalho, 23, was attacked while surfing off of Boa Viagem beach, Recife. He sustained injuries to his left calf and foot. Gleison Sena of the Pernambuco state fire department said that Carvalho had been surfing in a high-risk area, where there were ample warning signs to keep surfers away due to the frequent shark attacks.
- José Ivair Pereira, 35, had his leg bitten while swimming off Piedade, Pernambuco on 9 April 2006.
- On 21 August 2004, Wagner da Silva, 24, was bitten on his leg and hands while swimming off Boa Viagem Beach, Recife.
- Valmir Pereira Silva was attacked as she waded off Piedade Beach, Recife on 23 May 2004. She sustained wounds to her legs and buttocks.

- On 29 March 2004, Alcindo Sousa Jr., 23, lost a leg when swimming off Piedade Beach.
- Felipe Tavares Marinho, 16, was bitten on the finger in the waters off Copacabana Beach, Rio de Janeiro on 25 April 2003.
- On 23 April 2003, Tiago Augusto da Silva Machado, 17, sustained appalling injuries that resulted in his left leg being amputated following an attack while surfing off Pau Amarelo Beach in the Paulista District of Pernambuco.
- Fabricio Jose de Carvalho, 19, was swimming off Piedade Beach when grabbed by either a Bull or Tiger shark on 16 September 2002. His left thigh was bitten and his leg had to be amputated.
- On 10 July 2002, Mario Cesar, 25, had his right arm severed while surfing just 15m (49ft) from Praia de Boa Viagem. He said: 'The shark appeared under me. I tried to punch him, but he took my arm and pulled me into the water. Eventually, I wrenched myself free and a big wave pushed me in to shore.'
- Paulo Fernandes Alves Ferreira, 40, had his leg seized by a shark as he surfed off Pina Beach on 10 May 2002.
- On 26 December 1999, Walmir da Silva, 18, was attacked while swimming in water barely up to his waist. Today, he wears a prosthetic leg below the left knee and also has a prosthetic arm. He said: 'The shark pulled me under with so much power that I really thought I was going to die. And I was losing blood from my leg. But I hit his dorsal fin and kicked out, and in the end he released me.'

Out of all South America's shark attacks, 89 have occurred in Brazil's waters with 21 fatalities. This large number is blamed on the 1984 development of Porto Suape to the south of Recife, an industry handling more than four million tons of cargo a year. The building work sealed off two freshwater estuaries that had discharged into the Atlantic Ocean and served as birthing waters for many Bull sharks. They then headed for the next estuary, which discharges into Recife's waters and a nearby channel used by surfers became the sharks' new feeding ground.

Fabio Hazin, marine biologist and head of the state-funded monitoring committee, said: 'Female Bull sharks used to enter those estuaries to give birth. From when the port was built, we believed a number of females moved north to the next estuary, which discharges onto the stretch of beach where the attacks happened.' Based on that finding, local human rights lawyers began to consider whether the state of Pernambuco should give compensation to the victims of shark attacks. The Safe Beach Project also proposed a solution to Recife's shark problem: a call to the local authorities to partition off a short section of beach using heavy-duty nets out at sea. Electromagnetic buoys would also deter – but not kill – the sharks. These are similar to the measures in place in both South Africa and Australia, but the idea has met with some opposition in Recife.

Completing the shark paradise is a slaughterhouse which disposes of blood into the estuaries. In all, Recife's

20km (12.5-mile) coastline has become an extremely dangerous place, with a higher proportion of attacks resulting in death. One in three shark attacks that occur in Recife are fatal.

On 5 February 2007, hip travel guide Brazil/Max penned a piece on 'the revenge of the sharks' at Recife, saying: 'If the pickpockets don't get you, the sharks will.' In an open letter to surfers (the most frequent victims of shark attacks), marine geologist Luiz Lira, scientific director of the Oceanic Institute, put it a little more technically: 'Sharks are coming closer to the beach and have attacked surfers and swimmers not to satisfy their hunger with human meat – they are looking for what's left of the fish, turtles, octopuses, squid, lobsters and crabs. When the Pernambucan sea was fertile, with an abundant menu, encounters between sharks and bathers were sporadic.'

On 13 May 2000, around 20 or so circling Tiger and Bull sharks prevented the recovery of the bodies of three people who died when a Piper aircraft crashed into the sea at Mont Dore in the South Province of New Caledonia. Three months earlier, on 15 March 2000, Gilbert Bul Van Minh, 35, was fatally attacked by what was thought to be a Tiger shark while spearfishing at Poum in the North Province. There were two shark attacks in New Caledonia in 2007: Stephanie Belliard, 23, was killed while swimming in the Bay of Luengoni in the Loyalty Islands on 30 September, and on 25 January 2007, Jesse Jizdny, 30, had his leg badly bitten by a Tiger shark at Kaala-Gomen in the North Province.

On 1 February 2006, swimmer Tessa Horan, 24, died from a torn leg after being attacked by a Tiger shark at Tu'anuku, Vava'u, Tonga. Another swimmer, Felipe Tonga, fared a little better after swimming with humpback whales at the same location on 27 September 2002: he escaped with a lacerated thigh when a Tiger shark attacked.

SHARK PREDATORS OF PAPUA NEW GUINEA (AND OTHER PLACES, TOO!)

Located in the Pacific Ocean, Papua New Guinea has logged 49 shark attacks and 25 fatalities since 1925. The numbers just top New Zealand, which has seen 47 attacks and 9 fatalities since 1852. However, New Zealand has 15,134km (9,404 miles) of shoreline, while Papua New Guinea has but 5,151km (3,201 miles) of coastline.

The waters of Papua New Guinea contain a wide array of marine environments, so divers visitors from all over the world come to the island to witness the immense variety of aquatic life, with shark dives one of the popular options. It's not clear whether Papua New Guinea's shark attacks stem from divers and other tourists, or if they originate from the local habit of fishing for sharks. Fisheries in Papua New Guinea exported $1.2 million (over £750,000) in shark-fin products in 1999.

More traditional means of fishing still exist there as well and reflect the fact that sharks have always been a part of the natives' lives. Some residents of Papua New Guinea, particularly those in the province of New Ireland, still practise the ancient art of shark calling. Shark callers claim to commune with shark spirits, drawing them near through ritual songs and prayers. When the shark comes to the boat, the caller places a noose on it, then clubs it and takes it home for the villagers to eat.

Over the years, there have been several unconfirmed reports of shark attacks in Papua New Guinea, including a man known only as 'Pasinganlas' dying at Enuk Island, Kavieng, after a shark attacked his abdomen and leg; at Kerema on the Western Papuan Gulf, when another local had his hand bitten; a fatal shark attack on a man spearfishing at Kombe, Talasea, in the West New Britain Province; a fatal attack on a spearfisherman at nearby Bulu, who was savaged on his back and buttocks, and an attack on a 12-year-old boy, who suffered injuries to his legs and abdomen while swimming at Volupai, Talasea.

- On 12 September 2011, French diplomat Thomas Viot (30) was attacked and bitten on the leg by a shark as he kite-surfed near a reef at the country's capital of Port Moresby. Despite a wound that went down to the bone, causing a huge loss of blood, Viot managed to kite-surf back to a beach, where local people and friends rushed to his aid. Christian Lohberger said they were able to give first aid during the trip back to Port Moresby: 'He was wave-riding in big waves about a kilometre or so offshore when something knocked him off his board and when he surfaced, there was a bad cut on his right leg. The shark appeared to have launched itself out of the water at him to take his leg out. The injury was quite bad, with some deep wounds, but does not appear to be crippling in that not much flesh was actually ripped away.'

A professional kite-boarder, Viot was surfing with two doctor friends visiting from France – a plastic surgeon and a general practitioner. 'I don't know how I managed it after the attack, but somehow I succeeded in riding back to the shore with my kite surf. I was bleeding very badly,' he said, after being flown to the Queensland city of Brisbane for treatment. 'There are always plenty of sharks around the reef, but I didn't see this particular one which attacked me. It's the first time I've been attacked by a shark – and I hope it's the last.'

- A young boy was bitten on the left leg and ankle while standing in water off Long Island, Madang Province on 7 May 2000.

Fourteen crewmembers of cargo ship *MV Mia*, which sank in waters off Palawan Island in the western Philippines on 29 September 2007, were believed to have been eaten by sharks. Four survivors had heard the horror and were in deep shock when they were rescued. A local police spokesman said: 'The survivors said they heard the others screaming while asking for help. They said they saw their fellow seamen being eaten by the sharks but there was nothing they could do.'

Shark Encounters Happen in Britain, Too!
Fact: shark attacks also occur in less exotic climes. Andrew Rollo, 26, was bitten by a shark while surfing with friends in the mouth of the Spey River, Moray, Scotland on 28 November 2011. Rollo used his surfboard to fend off the lunging shark – believed to be a Porbeagle, ranging from 2.4–3m (8–10ft) long. 'The first thing I knew about it was when it was touching my leg – it was right there. It went round the front of me and then it went over to a friend, who was only three metres away. It made a sharp turn to me, at which point I got off,' he said.

The rare attack sparked an investigation, then on 1 August 2002, in the Blue Planet Aquarium in Ellesmere Port, Cheshire, Rob Bennett, 30, was bitten on the hand while diving with sharks as shocked visitors looked on. Bennett had been carrying out routine maintenance work in the Caribbean Tank when one of the Sand Tiger sharks took offence at the intrusion. A spokesman for the Aquarium said the incident had been dealt with quickly and that the injured man had been taken to hospital for treatment, adding: 'Rob is in good spirits but feeling a little embarrassed.' One visitor signing herself 'Concerned Parent' commented: 'I don't think those big sharks should be kept in captivity like that – if they are biting divers, they are obviously not happy.'

On 18 August 2011, a shark savaged a 16-year-old boy as he swam off Zheltukhin Island in Russia's Far East. It was the second attack in 24 hours. Earlier, Denis Udovenko, 25, lost both arms to the elbow when a shark attacked him in Telyakovsky Bay, near Vladivostok. Beaches in the area were closed and people were advised to stay out of the water. Advice was also issued on how to react in the event of a shark attack, saying victims should try and beat the beast away 'especially with strikes to the eyes and gills.'

Aleksandr Sokolovsky of the regional branch of the Russian Academy of Sciences said that sharks had been seen in the area for some time but added: 'In the 51 years that I've worked as an ichthyologist [zoologist specialising in fish] this is the first confirmed case of an attack on a human here.' The shark (or sharks) involved in the two attacks was likely to be a Great White. Some experts suggest the growing presence of sharks in the region might be linked to climate change with higher seawater temperatures attracting crabs, on which the sharks feed. Others said anchovies in warm shallow waters attracted sharks.

- On 19 June 2011, Kevin Moraga, 15, died after an attack while surfing off Zheltukhin Island, Puerto Rica. Tragically, it was first thought thc boy would survive his injuries but he died five days later in hospital.

ACKNOWLEDGEMENTS

- Dr. George Burgess, director of the Florida Program for Shark Research and curator of the International Shark Attack File at the Florida Museum of Natural History
- Ralph Collier, Shark Research Committee
- Marie Levine, head of the Shark Research Institute at Princeton
- Dr. Erich Ritter
- Samuel Gruber, head of Miami's Bimini Biological Field Station
- Dr. Elke Bojanowski
- Yannis Papastamtiou
- Dr. Jonathan Werry
- Jeremy Cliff
- Tim Ecott: *Neutral Buoyancy: Adventures in a Liquid World* (Penguin, 2002)
- FLMNH Ichthyology Department: International Shark Attack File
- Global Shark Attack File

ACKNOWLEDGEMENTS

- HEPCA
- The Complete File of South Africa Shark Attacks and Incidents
- Australian Shark Attack File
- Queensland Government
- Observatory Martine Reunion
- Seychelles Tourist Board
- Channel 5
- Sky News
- ABC News
- The *Sun*
- *National Geographic*
- *Sharm Business & Community Magazine*
- *San Francisco Chronicle*
- *Sunday Telegraph Magazine*
- *Perth Now*
- *Daily Mail*
- Dailymail.co.uk
- *Daily Telegraph*
- Telegraph.co.uk
- Jennifer Reade
- The Complete South African Shark Attack and Incident Record
- The Foreign and Commonwealth Office
- Shark Trust and Shark Conservation Society
- PETA
- Pew Environment Group
- Shark Attack Survivors
- Egypt Today
- *Travelweekly*
- ZigZag News

- Travel.yahoo.com
- Viktormar.wordpress.com
- Travelandleisure.com
- Brazil/Max
- Surfline
- Sharkattacks.com
- Hawaiisharkencounters.com
- Oceana
- Sharkattackfile
- Surftherenow.com
- Sharkresearchcommittee.com
- Thedivingblog.com
- Surfmag.com
- Surfinghandbook.com
- Sharkattacksurvivors.com
- Yosurfer.com
- The Equalizer
- TopKayaker.Net
- Haarezt.com
- The Art of Manliness